RAPID REVISION NOTES

A LEVEL

PHYSICS

Electricity
and Magnetism

RAPID REVISION NOTES

A LEVEL

PHYSICS

Electricity and Magnetism

BY F A GREEN

Ph.D.
Head of Physics,
Tunbridge Wells Technical High School
for Boys

General Editor — Professor A. J. B. Robertson,
Professor of Chemistry, King's College, London

CELTIC
REVISION AIDS

Celtic Revision Aids
30–32 Gray's Inn Road,
London WC1X 8JL

First Published 1982

ISBN 17 751 4655

Printed and bound in Great Britain by
Cox & Wyman Ltd, Reading

General Editor's Foreword

The Rapid Revision 'A' Level Series is designed for students preparing for G.C.E. 'A' Level, Scottish Highers, Intermediate University and similar examinations. They are comprehensive and may be used either on their own for revision or as a complement to set books and text books. The notes are organised to help students remember facts and to pin-point areas of difficulty. Practice questions are included at the end of each section with answers to the numerical problems at the end of the text. Where necessary, clearly worked through examples have been included in the text.

I am sure students will find these books very helpful.

A J B Robertson
M.A., Ph.D., D.Sc., C. Chem., F.R.S.C.

Professor of Chemistry,
King's College London,
University of London

Formerly Fellow of St John's College
Cambridge

Author's foreword

This book is designed for use alongside the student's own notes and text books, to provide a complete revision programme in this particular branch of the subject.

The book is divided so that each chapter deals with a main topic of the syllabus. Within each chapter there are concise definitions and notes, worked through examples and, finally, practice questions for the student to attempt. Answers to numerical questions are provided at the back.

F. A. Green

CONTENTS

1 BASIC IDEAS AND DEFINITIONS

The structure of matter

The idea of matter consisting of small almost indestructible particles called atoms first evolved amongst the ancient Greeks and has been refined since then by various scientists. The most significant contributions have, perhaps, been made by Dalton and Rutherford. Tracing this history is beyond the scope of these revision notes which concentrate on the main facts which have emerged.

1. Matter consists of **atoms** which are sometimes arranged in certain definite patterns (a crystal lattice).
2. The diameter of an atom is of order 3×10^{-10} m.
3. Almost all of the mass is concentrated in a small central **nucleus** of diameter about 10^{-14} m.
4. The nucleus contains two kinds of particle namely **protons** and **neutrons**.
5. The proton and neutron have approximately the same mass but whilst the neutron is uncharged the proton is positively charged.
6. The rest of the space (i.e. most of the atom) is occupied by **electrons** which are negatively charged and are in motion around the nucleus.
7. The proton and neutron are about 1840 times heavier than an electron (often approximated to a factor of 2000).
8. Since an atom is neutral, the number of electrons orbiting the nucleus is equal to the number of protons inside it.
9. If the atom gains or loses an electron it acquires a net charge and is then called an **ion**.
10. The atoms of different elements vary in that they have different numbers of protons, neutrons and electrons.

This is best represented in a table.

Element	**No. of protons**	**No. of neutrons**	**No. of electrons**
Hydrogen	1	0	1
Helium	2	2	2
Carbon	6	6	6
Nitrogen	7	7	7
Oxygen	8	8	8
Sodium	11	12	11

Electric current

Since matter consists of atoms it therefore consists of charged particles which may be free to move throughout the material. If the charges can be made to flow then an electric current is said to pass and the substance is termed a **conductor**. Substances in which no charges are free to move are called non-conductors or **insulators**. (There is a third class of material known as a semi-conductor which will be discussed later.)

The two main types of conductor are:

1 **Metals**
The atoms are fixed on permanent sites but the outermost electrons are shared with other atoms and are free to move.

2 **Electrolytes**
When some chemical compounds go into solution they break into positively and negatively charged ions which can move through the liquid and conduct an electric current.

Unit of current

Ampère (or amp) Symbol I

The amp is defined, somewhat inconveniently as being: 'That current which flowing in each of two infinitely long parallel straight wires of negligible cross-sectional area separated by a distance of one metre in a vacuum produces a force of 2×10^{-7} Newton metre^{-1} between them'.

This definition will be discussed again later in these notes.

Conventional current flow

Since charged particles would flow in one direction if they were positive but in the opposite direction if they were negative, it is necessary to adopt a sign convention for the direction of the electron current. The positive direction of the electric current at a particular point in a circuit is taken to be the direction in which a positive charge would flow if placed at that point. This direction is indicated by an arrow on the circuit diagram in Fig. 1 where a battery is shown driving a current through a resistor.

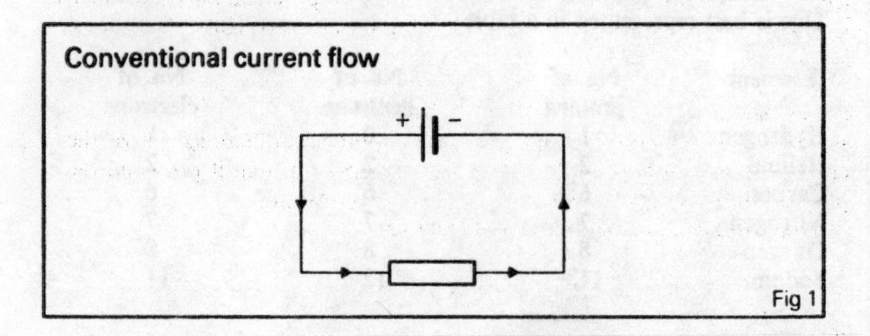

Fig 1

This convention requires careful interpretation since in Fig. 1 the current will, in fact, be carried by negatively charged electrons which will move in the opposite direction to that shown by the arrows.

Electron flow is in the **opposite direction** to **conventional current flow** which is shown on the circuit diagram.

Electric charge
Unit Coulomb Symbol Q

One coulomb is **defined** as being the charge which flows when a current of one amp flows past a given point for one second.

This can be expressed mathematically as:

$$Q = It$$

where t is the **time** in **seconds**

Potential difference
Unit volt Symbol V

In general a current will not flow in a circuit unless an energy source such as a battery is provided since the charges will be evenly distributed and therefore unable to move. The battery gives energy to the circuit by setting up an uneven charge distribution by transferring electrons internally from one pole to the other.

The battery is said to provide a source of **electromotive force** or **e.m.f.** and there is a **potential difference** or **p.d.** between its terminals.

The potential difference (p.d.) between two points is **defined** as being one volt if one joule of work needs to be done in taking one coulomb of positive charge from one point to the other.

This can be expressed mathematically as:

$$W = QV$$

where W is the work done in joules

It is usually more convenient to assign a potential to a particular point in a circuit by relating it to some reference point. For practical purposes earth is usually chosen as the reference and given the value zero volts. This choice is purely arbitrary though clearly convenient.

Ohm's law

For certain conductors (Ohmic devices) the potential difference across the conductor is proportional to the current flowing through it provided the physical conditions (e.g. temperature) remain constant.

This can be expressed mathematically as:

$$V \propto I$$

and hence

$$V = IR$$

where R is the constant of proportionality called the **resistance** of the material measured in ohms (Symbol Ω)

The following points should be noted:

1 Any ammeters or voltameters used in the verification of Ohm's law must have a calibration which does not itself depend on Ohm's law. Ordinary

moving coil ammeters and voltmeters are, therefore, not available and a valid experimental arrangement presents a difficult problem.

2 Ohmic conductors possess a linear V/I characteristic but not all devices exhibit this type of behaviour as shown in Fig. 2.

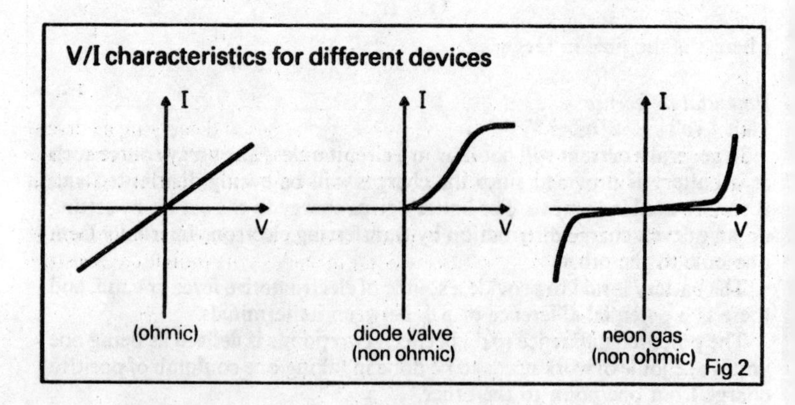

Fig 2

Electrical power

Unit Watt Symbol P

The electrical power is the energy released per second from the device **(1 watt = 1 joule per second)**. This energy is typically in the form of light or heat.

Since the amount of energy given out by a charge of Q coulombs in passing through a device which has V volts across it is given by:

$$W = QV$$

$$\therefore \quad W = ItV$$

and

$$P = \frac{W}{t} = IV$$

$$\textbf{power in watts = volts} \times \textbf{amps}$$

$$P = VI = V^2/R = I^2R$$

N.B. It is only necessary to remember one of these three power expressions since the other two can be evaluated when required by using Ohm's law.

Resistors in series and parallel

Any combination of resistors in series and parallel can be reduced to a single equivalent resistor to simplify circuit analysis.

a Series

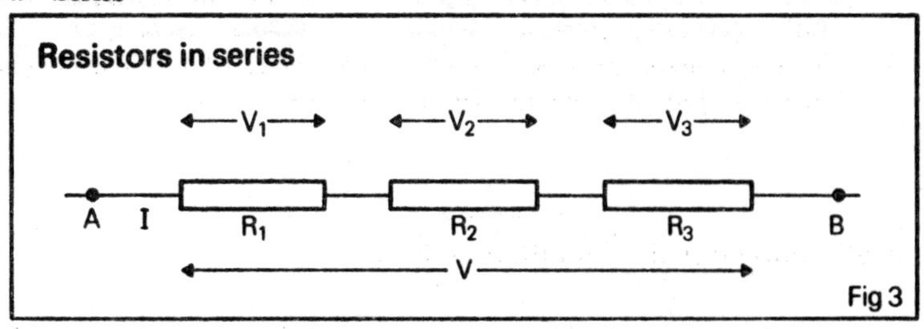

Fig 3

Consider three resistors $R_1R_2R_3$ connected in series and carrying a current I. Suppose that the resistors have potential differences $V_1V_2V_3$ across their ends. Let R be the total resistance and V the resultant potential equivalent resistor R.

Since the total energy supplied to the system per second is equal to the sum of the energies dissipated per second in each of the individual resistors we can therefore write:

$$IV = IV_1 + IV_2 + IV_3$$
$$\therefore\ V = V_1 + V_2 + V_3$$
$$\therefore\ IR = IR_1 + IR_2 + IR_3$$
$$\text{Hence } \mathbf{R} = \mathbf{R_1} + \mathbf{R_2} + \mathbf{R_3}$$

The following points should be noted for a combination of resistors in series:

1 The total resistance is the sum of the individual resistors and the **resistance is** therefore **increased when extra resistors are added in series.**
2 The same current passes through each resistor. (The current in a series circuit is everywhere the same. The current is **not** used up.)
3 The total potential difference is the sum of the individual potential differences which are proportional to the individual resistances. (The basic idea of the potential divider which is discussed later.)

b Parallel

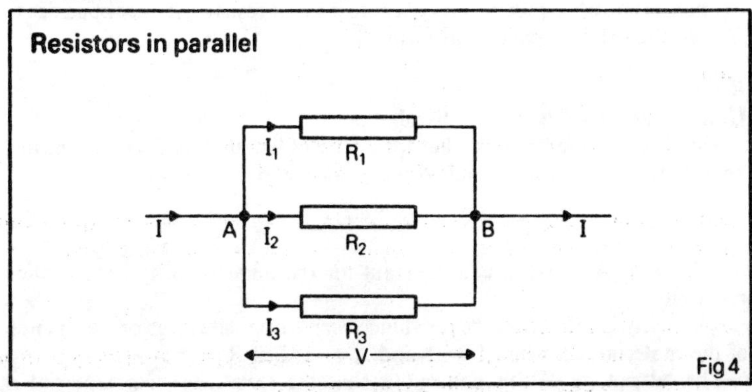

Fig 4

This analysis can be repeated for the parallel case where the resistors carry different currents $I_1 I_2 I_3$. The potential difference across each resistor is the same since they are all connected between the same two points A and B.

Since the current splits on reaching point A we can write:

$$I = I_1 + I_2 + I_3$$

$$\therefore \frac{V}{R} = \frac{V}{R_1} + \frac{V}{R_2} + \frac{V}{R_3}$$

where R is the single equivalent resistor

Hence $$\frac{1}{R} = \frac{1}{R_1} + \frac{1}{R_2} + \frac{1}{R_3}$$

The following points should be noted:

1 The effective resistance R is **always less** than any of the individual resistors $R_1 R_2 R_3$. **Adding resistors in parallel always reduces the total resistance.** A wire which is to carry a large current should therefore be as thick as possible to reduce the resistance and minimise the power loss in the cable.

2 In the **special case of two resistors only** in parallel the expression reduces to:

$$R = \frac{R_1 R_2}{(R_1 R_2)}$$

(the product of the two individual resistors divided by their sum)

$$R = \frac{\textbf{Product}}{\textbf{sum}}$$ for **two resistors only** in parallel

This expression is simpler than the original form since reciprocals are not involved. It can still be used if there are more than two resistors since the resistors can always be taken in pairs, each pair being reduced to a single equivalent resistor. However, since more than one calculation is now required to reduce the system to a single resistor, it may be easier to use the original reciprocal formula.

Resistivity

Unit Ohm metre (Ω m) Symbol ρ

Simple experiments show that for a wire of length ℓ and cross-sectional area A, the resistance depends directly on ℓ and inversely on A.

$$\therefore R \propto \ell/A$$

and $R = \rho\ell/A$ where ρ is a constant for the material of the wire called **resistivity**.

Resistivity is defined as the resistance across opposite faces of a unit cube of the material (i.e. when $\ell = 1$ and $A = 1$ then $R = \rho$).

The advantage of this concept is that the resistivity of a material is constant (provided the temperature remains constant) whilst resistance is not since the resistance is clearly a function of the dimensions. The resistivity of a material can therefore be incorporated into a data book whilst the resistance cannot.

N.B. Since $\rho = RA/\ell$ the units of ρ are ohm metre (Ω m) and **not** ohm metre^{-1}. This is a common mistake.

Conduction in a metal and drift velocity

Electrical conduction in a metal is due to free electrons which wander randomly from atom to atom in the absence of an externally applied electrical field. When a battery is connected across the ends of the metal an electric field is established and the electrons are accelerated by this field and move in a given direction. Collision with atoms vibrating on their lattice sites causes a loss of electron energy (and an increase in metal temperature) but the electrons are then accelerated once again by the electric field and the process repeats itself. The movement of electrons in the field direction is, therefore, erratic **but on average they progress with a mean velocity called the drift velocity** which can be calculated.

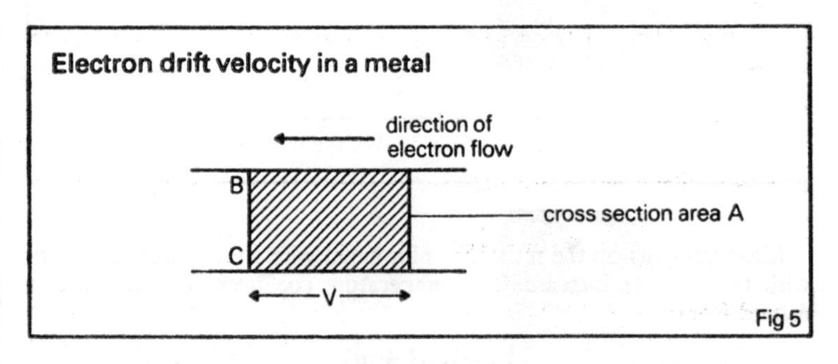

Suppose a wire has cross sectional area A and contains n free electrons per cubic metre each travelling with a drift velocity of V metres per second.

In one second, all of the electrons in the shaded region (volume AV) will pass the line BC.

$\therefore$ Charge passed per second = n AVe where e is the electronic charge

But since current (I) is the charge passed per second we can write:

$$I = nAVe$$

Inserting some typical values for copper (I = 10 A, A = 1 mm^2 (10^{-6} m^2), e = $1{\cdot}6 \times 10^{-19}$ C and n = 10^{28} electrons m^{-3}) and rearranging the above equation gives a figure for V of order millimetres per second. This value is, perhaps, surprisingly low but explains why the term 'drift velocity' is used.

N.B. Since the drift velocity is so low the bulk motion of electrons cannot be responsible for the sending of electrical signals along a wire. An electrical impulse can be transmitted along a cable from England to America in about

1/100 second yet an individual electron would take approximately 300 years to cross the Atlantic Ocean.

The movement of an electron at one end causes the similar movement of another electron at the other end almost immediately since the forces which influence electrons are propagated at approximately the speed of light (3×10^8 m s^{-1}).

Temperature coefficient of resistance (TCR)
Unit K^{-1} (per Kelvin) Symbol α
or °C^{-1} (per degree centigrade)

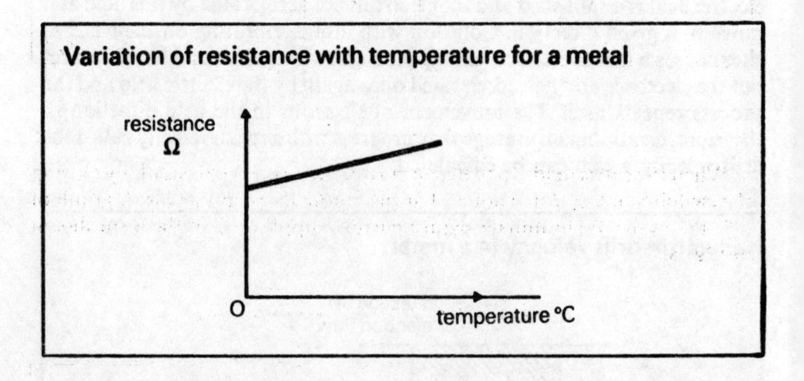

Measurements on the resistance of a metal wire show that it increases with temperature increase. The temperature coefficient of resistance is **defined** from:

$$R_t = R_0(1 + \alpha t)$$

where R_t is the resistance of temp t
R_0 is the resistance at 0°C
t is the temperature

Rearranging gives $\alpha = \dfrac{R_t - R_0}{R_0 t}$ and enables the TCR to be **defined**.

The temperature coefficient of resistance of a material is the increase in resistance per degree C rise of temperature divided by the resistance at 0°C.

N.B. The definition is related to the resistance at 0°C (see example at the end of this chapter).

Internal resistance of a cell
When a voltmeter is connected across a cell the e.m.f. of the cell can be measured and can indicate how many joules of energy are released for each coulomb of charge which passes.

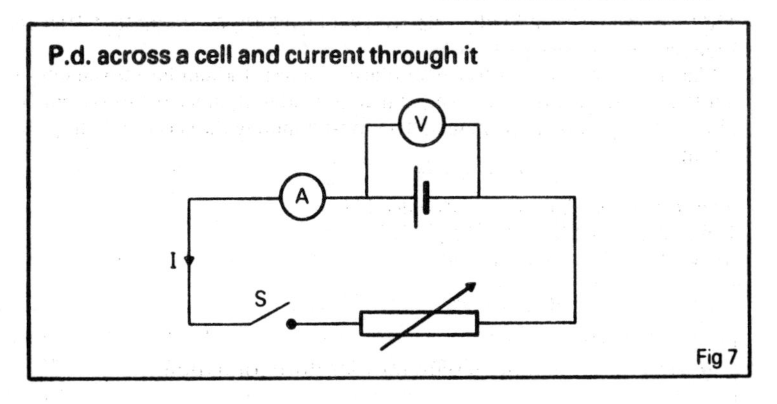

When the cell is made to deliver a current to an external circuit by closing the switch S in Fig. 7 it is noted that the voltmeter reading falls. A graph of p.d. across the cell terminals against current supplied takes the form shown in Fig. 8.

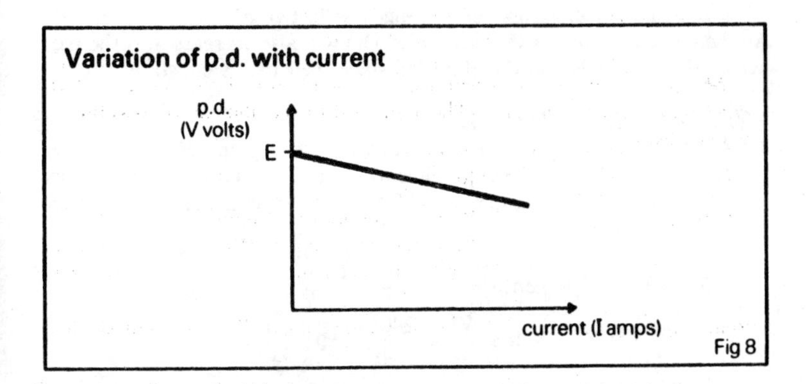

As the current increases the 'lost volts' also increase as the voltmeter reading continues to fall. These results can be explained by assuming that the cell itself possesses internal resistance inherent in its construction and that some energy is therefore used in driving the current through the cell itself. This energy is consequently not available to the external circuit and so the voltmeter reading falls.

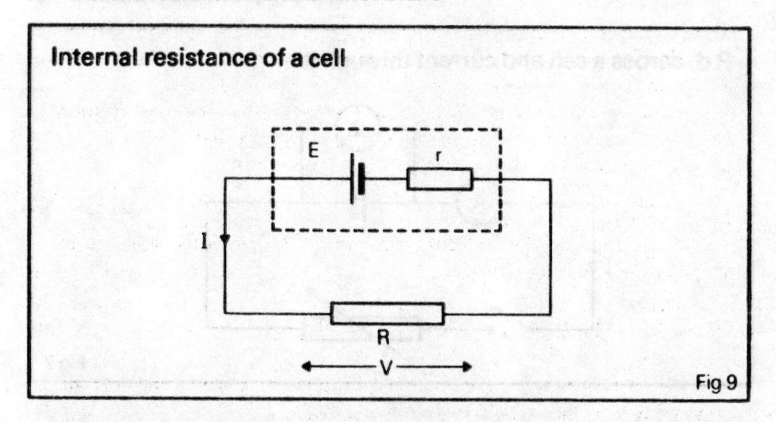

Fig 9

Consider a cell of e.m.f. E volts and internal resistance r ohms delivering a current I amps through an external resistor R ohms. Suppose that the potential difference across the terminals of the cell is V volts where $E > V$.

Applying Ohm's law to the external circuit: $V = IR$

Since the single equivalent resistor is $(R + r)$ ohms we can write for the complete circuit:

$$E = I(R + r)$$

Hence

$$E = V + Ir$$

e.m.f. = p.d. across cell terminals + 'lost volts'

As the current increases the 'lost volts' (Ir) will also increase and the p.d. across the terminals will therefore fall since the e.m.f. is constant.

Worked examples on chapter 1

1 A heating element is required which will dissipate 720 watts when connected to a 240 volt supply. The wire available to wind the heater has a diameter of 0·5 mm and resistivity $1{\cdot}0 \times 10^{-6}$ ohm metre. How much wire is required?

Since $P = V^2/R$ hence:

$$R = \frac{V^2}{P} = \frac{240^2}{720} = 80\,\Omega$$

Since $R = \rho\ell/A$ and $A = \frac{\pi d^2}{4}$ where d is the diameter

$$\therefore\ \rho = \frac{RA}{\ell} = \frac{R\pi d^2}{4} = \frac{80 \times (5 \times 10^{-4})^2}{4 \times 1 \times 10^{-6}} = 15{\cdot}7\text{ m}$$

This answer is rather high, suggesting that in practice some thinner wire would be an advantage.

2 Two cells X and Y each of e.m.f. 1·1 volt are connected in series with a

10·0 Ω resistor. A high resistance voltmeter reads 0·9 volt when connected across X and 1·0 volt when connected across Y. What is the internal resistance of the cells?

Since the voltmeter has a high resistance we can assume that it does not change the circuit characteristics when it is inserted.
The p.d. across both cells = 0·9 + 1·0 = 1·9 volts

The p.d. across both cells = 0·9 + 1·0 = 1·9 volts
∴ The p.d. across the external resistance is 1·9 volts since it is connected across both cells

$$\text{Since } V = IR \text{ hence } I = \frac{V}{R} = \frac{1 \cdot 9}{10 \cdot 0} = 0 \cdot 19 \text{ A}$$

Since the current in a series circuit is everywhere the same, the current through each of the cells is therefore 0·19 A.
Since E = V + Ir

∴ For cell X 1·1 = 0·9 + 0·19 r

$$\therefore \ r = \frac{1 \cdot 1 - 0 \cdot 9}{0 \cdot 19} = \frac{0 \cdot 2}{0 \cdot 19} = 1 \cdot 05\ \Omega$$

For cell Y 1·1 = 1·0 = 0·19 r

$$\therefore \ r = \frac{1 \cdot 1 - 1 \cdot 0}{0 \cdot 19} = \frac{0 \cdot 1}{0 \cdot 19} = 0 \cdot 53\ \Omega$$

3 A coil has resistance 76·0 ohms at 25°C. What is its resistance at 75°C if the temperature coefficient of resistance is $4 \cdot 0 \times 10^{-3}\ °C^{-1}$?

$$R_{25} = R_0\,(1 + 25\alpha)$$
$$R_{75} = R_0\,(1 + 75\alpha)$$

Dividing one equation by the other eliminates R_0 and produces:

$$\frac{R_{75}}{R_{25}} = \frac{1 + 75\alpha}{1 + 25\alpha} = \frac{1 + 75 \times 4 \times 10^{-3}}{1 + 25 \times 4 \times 10^{-3}} = \frac{1 \cdot 3}{1 \cdot 1}$$

$$\therefore \ R_{75} = R_{25} - \frac{1 \cdot 3}{1 \cdot 1} = \frac{65 \times 1 \cdot 3}{1 \cdot 1} = 76 \cdot 8\ \Omega$$

A quicker solution could be obtained by using an approximate formula:
Resistance increase = Initial resistance × α × temperature rise
∴ Resistance increase = $65 \times 4 \times 10^{-3} \times (75 - 25) = 13\ \Omega$

$$\therefore \ R_{75} = 65 + 13 = 75\ \Omega$$

The error made is, therefore, significant but it might be acceptable in applications where accuracy is not essential.

4 The electrons in a television tube have a speed of $8 \times 10^{7}\ \text{m s}^{-1}$ and travel 40 cm before hitting the screen. If the beam current is 3·0

milliamps, how many electrons are present in the beam at any one time? (Electronic charge $= 1{\cdot}6 \times 10^{-19}$ coulomb.)

$$\text{Time to cross the tube} = \frac{\text{Distance}}{\text{Speed}}$$

$$= \frac{0{\cdot}4}{8 \times 10^{7}} = 5 \times 10^{-9}\ \text{s}$$

In this time, all the electrons in the beam will reach the screen since $Q = 3 \times 10^{-3} \times 5 \times 10^{-9} = 15 \times 10^{-12}$ coulombs.
But the total charge arriving is given by:
Total charge = number of electrons in the beam × charge on one of them

$\therefore\ Q = n \times 1{\cdot}6 \times 10^{-19}$ when n is the total number
$\therefore\ 15 \times 10^{-12} = n \times 1{\cdot}6 \times 10^{-19}$

and $$n = \frac{15 \times 10^{-12}}{1{\cdot}6 \times 10^{-19}} = 9{\cdot}37 \times 10^{7}\ \text{electrons}$$

Practice questions

1 A battery of e.m.f. 2·0 volts and internal resistance 0·5 Ω is used to supply a current of 0·2 A through a lamp. If the lamp is to be run from a battery of e.m.f. 3·0 volts and negligible internal resistance, calculate the series resistor required such that the brightness of the lamp remains the same.

2 A 3·00 ohm resistor is required for a particular application but the only standard resistor available has a resistance of 3·05 ohms. Explain how a wire of resistance 10 ohms per metre can be used to modify the resistor to the value needed.

3 Explain the difference between the e.m.f. of a battery and the potential difference across its terminals. A battery has an e.m.f. of 4·0 volts but when a voltmeter is connected across its terminals the reading is only 3·8 volts. When a 10 ohm resistor is connected across the battery (with the voltmeter removed) the current is 0·3 A. Calculate the resistance of the voltmeter.

4 A wire has resistance 10·5 ohms at 20°C and 15 ohms at 100°C. Calculate a value for the temperature coefficient of resistance.

2 BASIC ELECTROMAGNETISM

When a current is passed along a wire there is a magnetic field associated with it. This can easily be demonstrated by the use of iron filings or a plotting compass. The **direction of the magnetic field** is taken as the direction in which a **north pole would move** if placed in the field (i.e. field line direction arrows point from a north to a south pole) and can be found using the right hand grip rule.

Right hand grip rule

If the wire is held in the right hand with the thumb pointing in the direction of the current, then the direction in which the fingers wrap round the wire shows the direction of the magnetic field.

Magnetic field of a long straight conductor

The magnetic field near a long straight wire is shown in Fig. 10 where the concentric circles represent the field direction.

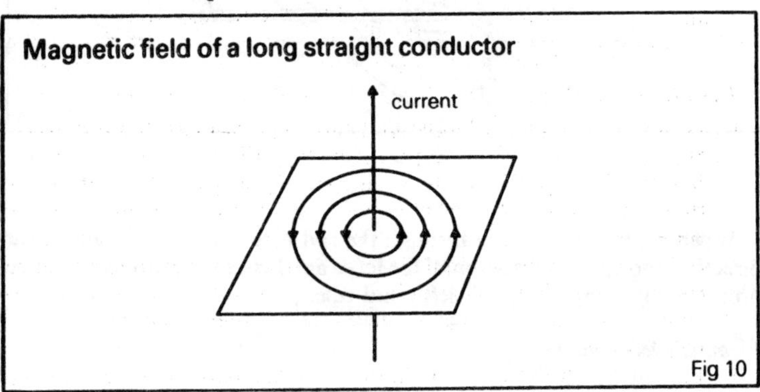

Fig 10

The information is perhaps better represented with the dot and cross notation shown in Fig. 11 in which current flowing into the paper is shown by a cross (tail of the disappearing arrow) and current flowing out of the paper by a dot (point of the approaching arrow).

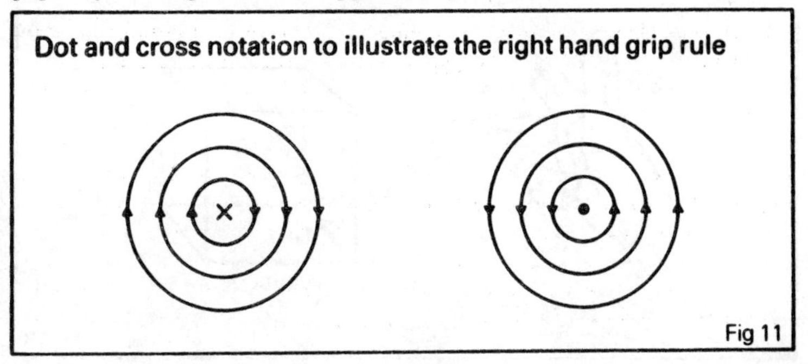

Fig 11

Force on a conductor in a magnetic field

When a current carrying conductor is placed in an external magnetic field (produced by a permanent magnet for example) it experiences a force and will move if free to do so. This can be demonstrated using the apparatus shown in Fig. 12 where a metal rod is placed across metal rails which lie between the poles of a horse shoe magnet.

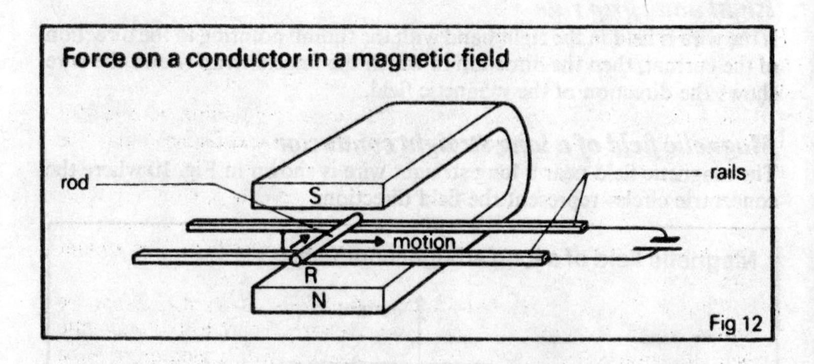

Fig 12

When a current is passed through the rod it rolls along the rails in the direction shown. The direction of the force and hence the movement can be predicted by using Fleming's left hand rule.

Fleming's left hand rule

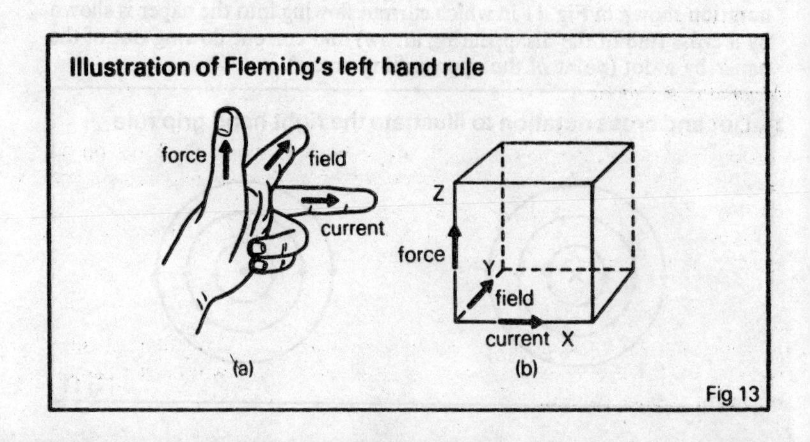

Fig 13

If the thumb, first finger and second finger of the left hand are held mutually perpendicular (at right angles to each other) with the first finger in the magnetic field direction and second finger in the current direction, then the direction of the thumb will indicate the direction of motion (or force).

Field – First finger
Current – Second finger
Motion – Thumb

Factors affecting the size of the force

Simple experiments show that the magnitude of the force (F) on the wire depends on four basic factors:

1 The magnetic field strength often called the magnetic flux density (B). The greater the field strength the greater the force experienced and F is proportional to B ($F \propto B$).
2 The size of the current (I). The greater the current the greater the force and $F \propto I$.
3 The length of the conductor (ℓ) which is placed in the field. The greater this length the greater the force and $F \propto \ell$.
4 The angle θ between the conductor and the field direction. It can be shown that $F \propto \sin\theta$ and hence a conductor parallel to the magnetic field direction experiences no force ($\sin 90° = 0$) whilst maximum force is observed when $\theta = 90°$ ($\sin\theta = 1$) and the conductor is at right angles to the field.

Combining these four factors together gives:

$$F \propto BI\ell \sin\theta$$

Hence $F = KBI\ell \sin\theta$ where K is a constant

This expression is used to define magnetic flux density (B) on the S.I. system.

Magnetic flux density

Unit Telsa (T) Symbol B

One telsa is the flux density in a magnetic field when the force on a conductor one metre long placed perpendicular to the field and carrying a current of one amp is one newton.

Hence when $F = 1$, $B = 1$, $\ell = 1$, $\sin\theta = 1$, and so $K = 1$ on the S.I. system. The expression therefore simplifies to:

$$\mathbf{F = BI\ell \sin\theta}$$

The magnetic flux density (B) is expressed in telsa (T) or alternatively weber metre^{-2} (Wbm^{-2}) where $1T \equiv 1\ Wbm^{-2}$. This will be discussed further in a later chapter.

If the wire is ℓ metres long and the drift velocity of the charge is V metres per second we can write:

$\ell = Vt$ (Distance = velocity × time)
$Q = It$ (Relationship between charge and current)

Hence $$t = \ell/V = Q/I$$

and so $$I\ell = QV$$

An alternative expression can therefore be obtained by replacing $I\ell$ by QV and writing:

$$F = BQV \sin\theta$$

Calculation of magnetic flux density for different conductor configurations
The magnetic flux density (B) for any conductor shape can be evaluated using the **Biot Savart law.**

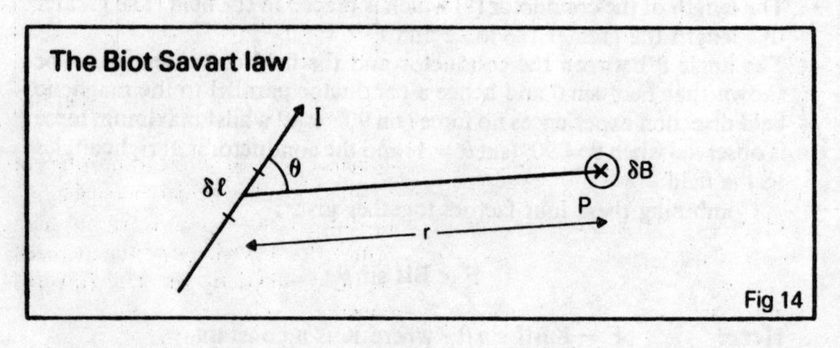

Fig 14

Consider a wire carrying a current I amps; the flux density at some point P due to a small segment of wire of length $\delta\ell$, a distance r away, is given by:

$$\delta B \propto \frac{I\delta\ell \sin\theta}{r^2}$$

where δB indicates that we have the small increment of flux density caused by the length $\delta\ell$.

This mathematical statement of the Biot Savart law cannot be proved directly since an infinitely small conductor of length $\delta\ell$ cannot be achieved in practice, but extending the theory to large scale practical configurations leads to predictions which can be verified by experiment.

The constant of proportionality depends on the medium between the wire and the point P but in the normal case of air (more strictly, vacuum) the law is expressed as:

$$\delta B = \frac{\mu_0}{4\pi} \frac{I\delta\ell \sin\theta}{r^2} \quad \textbf{Biot Savart law}$$

The constant μ_0 has the value $4\pi \times 10^{-7}$ Henry per metre ($H\,m^{-1}$) and is called the **permeability of free space** (the magnetic space constant). (The Henry as a unit of inductance is discussed in a later chapter.)

The magnetic field strength (B) at the point P caused by all of the wire can be calculated by summing all of the increments δB from each of the constituent lengths $\delta \ell$ of the wire. This process usually requires integration [$B = \int dB$] and is beyond the scope of these notes where we concentrate simply on the results from standard configurations.

Field at the centre of a circular coil of N turns

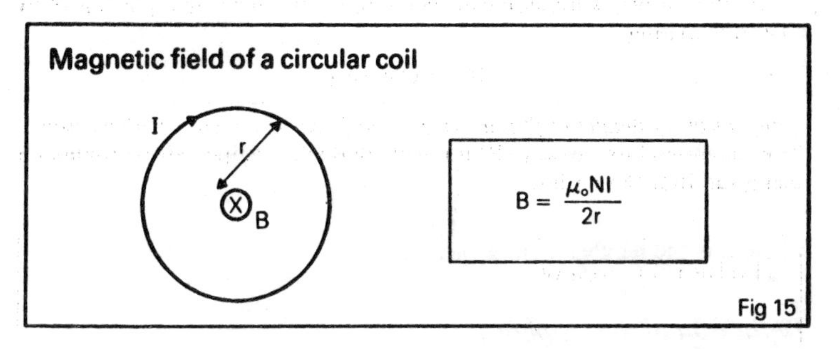

Fig 15

This shows the expected result that increasing the current and number of turns **increases** the **magnetic field strength** (flux density), whilst **increasing the coil radius** (r) **reduces** the **field strength** (although increasing the radius also increases the length of wire this is only a linear relationship and the increase in magnetic field strength is more than outweighed by the $1/r^2$ (inverse square) fall off caused by distance increase).

Field near a long straight wire

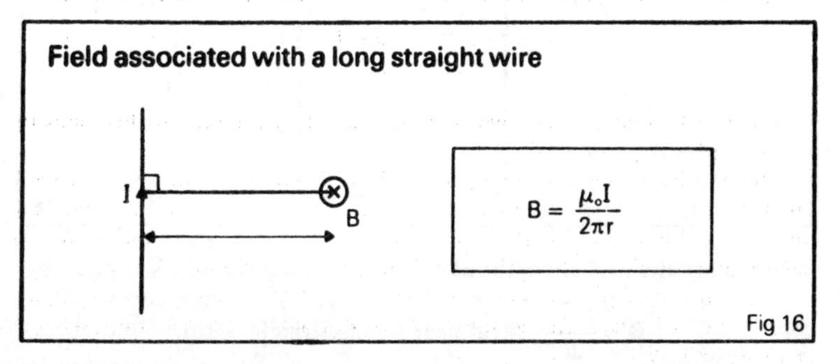

Fig 16

Field inside a solenoid

A solenoid is a long cylindrical coil which has a magnetic field configuration similar to that of a bar magnet.

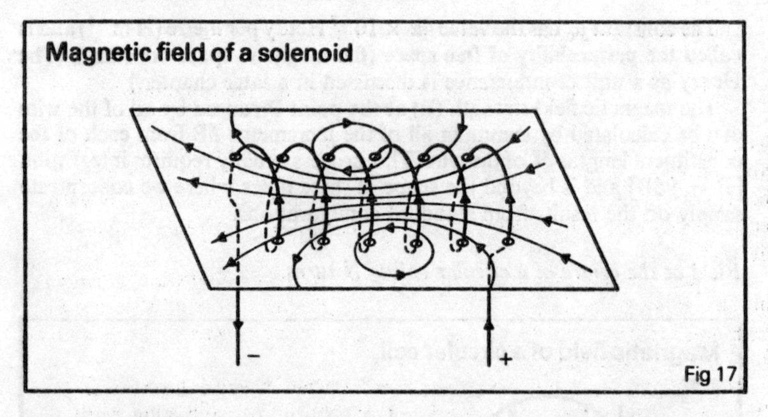

The magnetic field strength inside the coil is approximately constant and given by:

$$B = \mu_0 nI$$

where n is the **number of turns per metre**

Definition of the ampère

This definition was mentioned briefly in chapter 1 but is now considered in greater detail.

Consider two long straight parallel wires X and Y a distance r apart carrying currents I_1 and I_2 respectively.

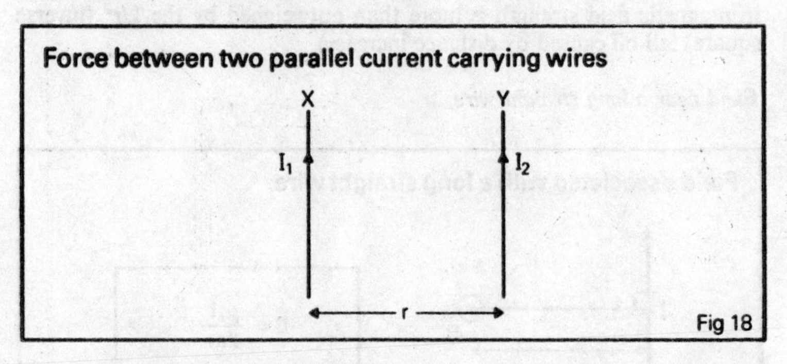

The magnetic field strength (B) at Y due to the current in X is given by:

$$B = \frac{\mu_0 I_1}{2r\pi} \quad \text{(field near a long straight wire)}$$

Y therefore experiences a force (F) given by:

$$F = BI_2\ell \quad \text{where } \ell \text{ is the length of the wire}$$

The force per unit length ($\ell = 1$) is therefore given by:

$$F = \frac{\mu_0 I_1 I_2}{2\pi r} \quad \text{in a direction towards X in the diagram shown}$$

The same argument is true for the magnetic field at X due to the current in Y and the direction of the force depends on whether the current in the wires is in the same or opposite directions. The wires are **attracted** to each other with **currents** flowing in the **same direction** but **repelled** with currents in **opposite directions**.

As stated previously, the ampère is defined as that current which flowing in each of two infinitely long parallel straight wires of negligible cross-sectional area, separated by a distance of one metre in vacuo, produces a force between the wires of $2 \times 10^{-7}\,\text{Nm}^{-1}$.

This definition is not particularly convenient since a force of $2 \times 10^{-7}\,\text{Nm}^{-1}$ may seem an unusual factor but it is consistent with a permeability of free space (μ_0) value of $4\pi \times 10^{-7}\,\text{Nm}^{-1}$.

The simple current balance

The force between two current carrying conductors a known distance apart is counter balanced by a known weight on a scale pan.

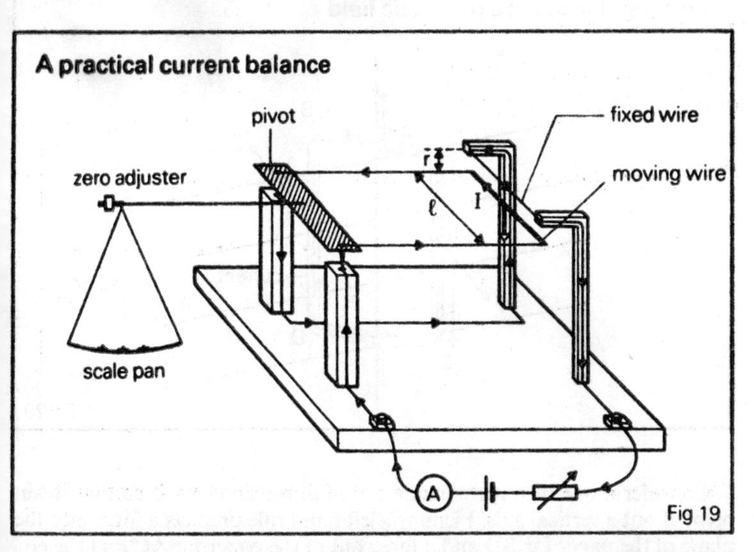

Fig 19

The two wires are arranged in series so that they carry the same current. Whilst one is fixed the other is free to move since the frame is pivoted as shown on the diagram. Before the current is switched on the zero adjuster is moved until the frame is horizontal. When the circuit is completed the moving wire is repelled downwards since the currents in the wires are in opposite directions. The balance is restored by adding a mass m to the scale pan.

The force between the wires is $\mu_0 I^2 \ell / 2\pi r$ and since the two arms of the balance are of equal length we can therefore write:

$$\frac{\mu_0 I^2 \ell}{2\pi r} = mg$$

where g is the acceleration due to gravity

Since all of the quantities can be easily measured the current can be found by this **absolute method** which does not rely on Ohm's law.

The moving coil galvanometer

The most convenient method of measuring current is achieved using a coil of wire suspended in a magnetic field.

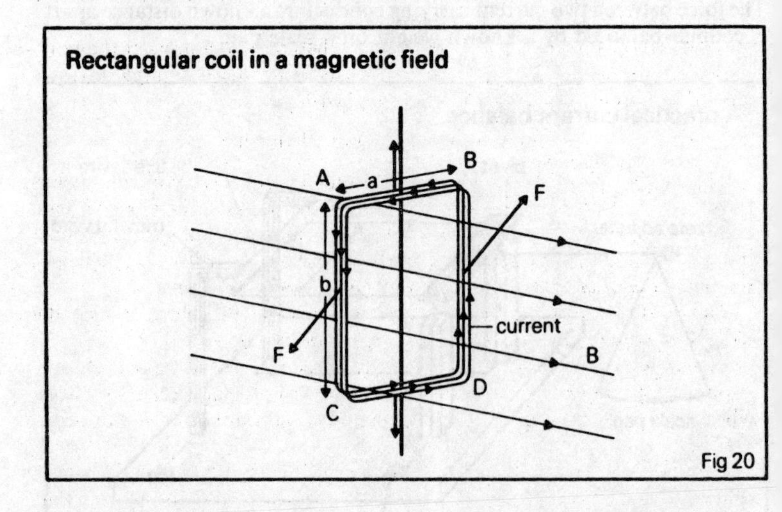

Fig 20

Consider a one turn rectangular coil of dimensions a × b so that it can pivot about a vertical axis. Fleming's left hand rule predicts a force into the plane of the paper for BD and a force out of the paper for AC and the coil therefore experiences a couple or turning effect. The coil will therefore rotate until the axis is parallel to the magnetic field direction when there is no turning effect. This is perhaps best illustrated with a view as seen from the top of the coil.

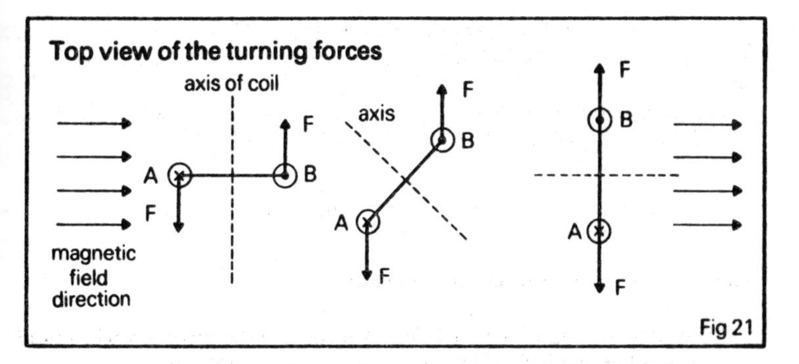

Fig 21

The force F on the vertical sides is BIb whilst any force on the horizontal sides need not be considered since it produces no turning effect.

Since the moment of a couple is given by:

Moment of couple = one of the forces × perpendicular distance between them

$\therefore$ Moment = Fa sin ϕ where ϕ is the angle between the axis of the coil and the external magnetic field

$\therefore$ Moment = BIba sin ϕ = BAI sin ϕ where A is the area of the coil

Since a practical coil will require n turns rather than a single turn we should now scale up this expression and write:

Moment of couple = C = BAnI sin ϕ

This simple coil is not suitable as an ammeter since the coil will always rotate to the same position irrespective of the applied current. We can, however, modify it by using either jewelled bearings and hair springs (robust but less sensitive) or a phosphor-bronze suspension (sensitive but delicate) so that a restoring torque (C_1) is applied as the coil rotates.

The restoring force is proportional to the twist (θ) which is also the deflection of the coil and home: hence not home.

Restoring couple = C_1 = Kθ where K is the suspension constant

As the coil rotates the restoring force increases until rotation finally stops when the deflection couple is equal to the restoring couple and we can write:

$C = C_1$ and so BAnI sin ϕ = Kθ

This expression can be simplified in practice since the magnetic field is shaped so that it always acts radially inwards towards the coil ensuring that ϕ is always 90° and the couple produced is a maximum. The radial field is achieved by using concave soft iron pole pieces for the magnet with a soft iron cylinder fixed centrally in the space inside the coil which is free to move in the resulting narrow cylindrical gap.

This radial field approximation will set sin ϕ = 1 and modify the expression to:

$$C = C_1 \quad \text{hence BAnI} = K\theta$$

$$\therefore\ I = \frac{K\theta}{BAn}$$

The deflecting couple is proportional to the current causing the deflection provided BAn are all kept constant and hence the instrument possesses the advantage of a linear scale ($\theta \propto I$) which is simple to read.

Sensitivity of the moving coil galvanometer

For a sensitive instrument a large deflection (θ) must be produced by a small current (I) and the current sensitivity (θ/I) needs to be fairly large.

$$\text{Current sensitivity} = \frac{\theta}{I} = \frac{BAn}{K}$$

Hence for high current sensitivity we require large values for BAn and a low value for the suspension constant K. These requirements are not all compatible since large values of A and n suggest a large coil whilst large B suggests keeping the gap between the pole pieces as small as possible and in practice a compromise has to be achieved. Similarly a large coil will be difficult to support from a fragile suspension wire which would be needed to achieve a low K.

The voltage sensitivity (θ/V) can be related to the current sensitivity since $V = IR$ (Ohm's law) where R is the resistance

$$\textbf{voltage sensitivity} = \frac{\theta}{V} = \frac{BAn}{KR}$$

The ballistic galvanometer

The ballistic galvanometer is intended to measure current pulses rather than steady currents. A galvanometer which can be used ballistically has a heavier coil than the normal galvanometer and as little damping as possible so that it will swing freely once started into motion. If the impulse is brief, the coil will immediately swing back towards the zero position, overshoot and continue oscillating backwards and forwards as energy is stored as potential energy in the suspension fibre and released again as it untwists. Normally a galvanometer has some built in damping to prevent these oscillations, but in the ballistic case (no damping) a number of oscillations will result before the motion is eventually dampened out by air friction and the internal friction of the fibre.

If the current pulse is short compared with the oscillation period of the coil, it can be shown that the amplitude of oscillation depends on Q (the total charge passed) and not on the way in which the charge varies during the pulse.

i.e. $Q \propto \theta$ where θ is the deflection of the coil

hence $Q = K\theta$ where K is a constant

The value of K can be found by calibrating the galvanometer by, for example, charging a known capacitor to a known voltage and then discharging it through the galvanometer. This is discussed in chapter 6.

In practice there are difficulties since some damping will always be present and we can only measure θ_1, the observed deflection on the first throw of the coil, whilst we require θ, the deflection in the absence of damping. $\theta > \theta_1$ but the coil never reaches a deflection θ because of damping.

If θ_2 is the magnitude of the second swing on the same side of zero then the damping factor on one swing is $(\theta_1 - \theta_2)$. Assuming that the damping is always constant, then the amount of damping in reaching θ_1 will be $(\theta_1 - \theta_2)/4$ since θ_1 is one quarter of one cycle from the start of the motion.

The true value of θ is therefore given to a reasonable approximation by:

$$\theta = \theta_1 + \left(\frac{\theta_1 - \theta_2}{4}\right)$$

Electric motor

The principle of the moving coil galvanometer can be extended to explain the operation of an electric motor. Here we discuss only a very brief outline of a complex topic which is studied in detail in many standard text books.

Consider again a current carrying coil in a magnetic field.

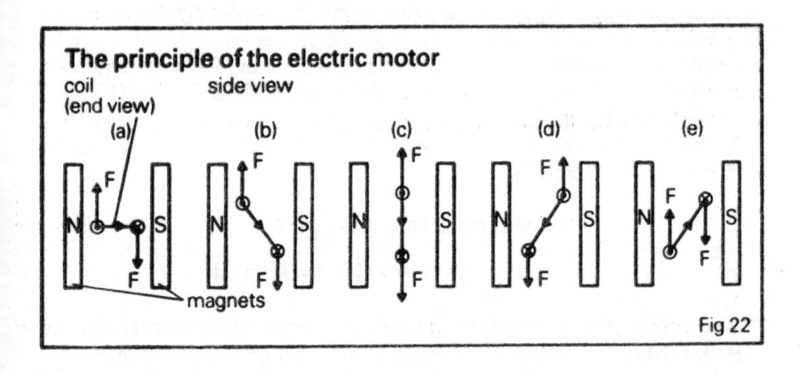

Fig 22

Diagrams (a) and (b) of Fig. 22, show the coil experiencing a couple and rotating but in (c) with the coil vertical the couple has reduced to zero and there is no turning effect. The momentum of the coil will, however, carry it over to position (d) but the couple causing rotation is now in the opposite direction and the coil will return to (c). For continuous rotation to be achieved the current direction in the coil needs to be reversed as shown in (e) when the rotating motion is able to continue.

This reversal of current direction can be achieved in two possible ways:

1 **D.C. motor.** This uses a **split ring commutator** (Fig. 23a) which reverses the connections twice every revolution. Contact is made with the rotating commutator via spring loaded brushes.

2 **A.C. motor.** An A.C. current input is given to the coil via slip rings and spring loaded contacts. (Fig. 23b.)

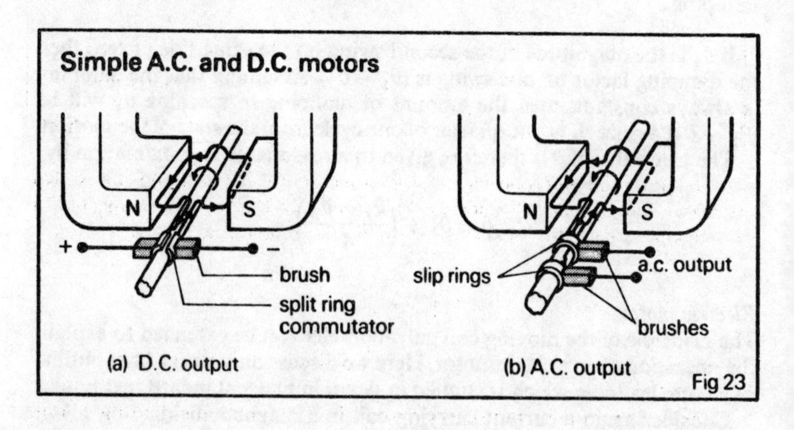

Ammeters and voltmeters

The basic moving coil galvanometer movement as previously described can be used either as a voltmeter or as an ammeter. Let us consider a specific example and suppose that the coil has a resistance of 30 Ω and requires a current of 10 milliamperes (mA) to produce a full scale deflection (FSD) of the pointer on the scale.

When a full scale deflection is achieved the potential difference across the coil is given by Ohm's law as:

$$V = IR = 30 \times 10 \times 10^{-3}$$

$$= 300 \times 10^{-3} = 0{\cdot}3 \text{ volt}$$

The meter is, therefore, already available for use as an ammeter reading to 10 mA (FSD) or a voltmeter reading to 0·3 volt (FSD) and would only require another scale to be painted onto its face so that reading the potential difference would be easier.

a Conversion to an ammeter

Now suppose that we wish to use the above meter movement to read currents up to one amp. Since the coil can only handle 10 mA (0·01 A) the other 990 mA (0·99 A) will need to be diverted round the meter movement by using a low resistance shunt (s) placed across the terminals.

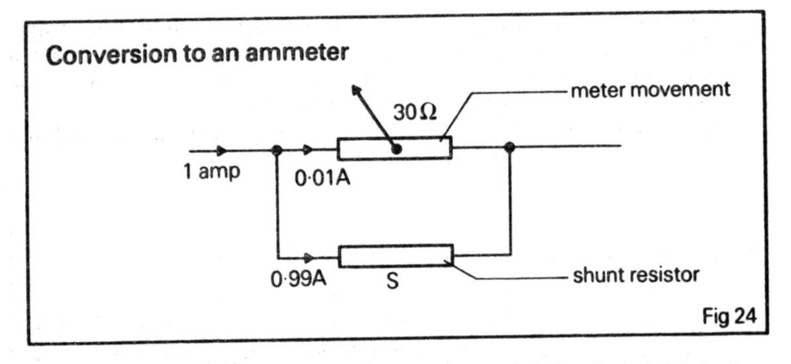

Fig 24

The potential difference across the movement is 0·3 volt and this is also the potential difference across the shunt since the two are connected in parallel. The value of the shunt is therefore given by:

$$S = \frac{V}{I} = \frac{0{\cdot}3}{0{\cdot}99}$$

$$= 0{\cdot}303\,\Omega$$

The meter can therefore be converted to an ammeter reading 1 amp (FSD) by connecting 0·303 Ω in parallel across its terminals.

b Conversion to a voltmeter

Now suppose that the same meter movement is to be used as a voltmeter reading to 10 volts.

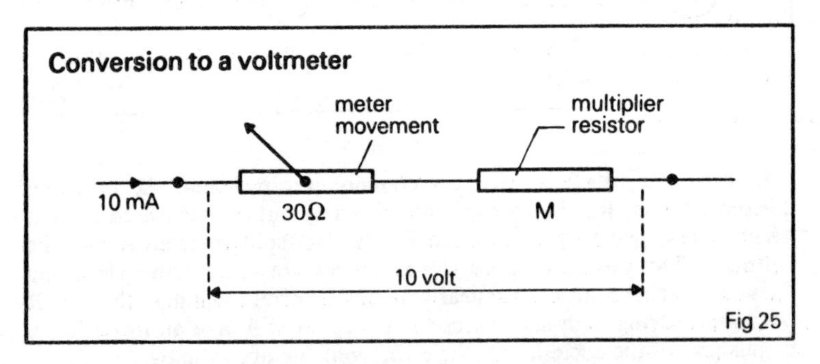

Fig 25

We now need to insert a multiplier resistor (M) in series with the coil such that there is a potential difference of 10 volts across the coil and multiplier when a current of 10 mA is flowing through them. Since the coil has 0·3 volt across it, as calculated earlier, the multiplier will have the remaining 9·7 volts across it with 10 mA flowing through it. The value of the multiplier resistor is therefore given by:

$$M = \frac{V}{I} = \frac{9{\cdot}7}{10 \times 10^{-3}} = 970\,\Omega$$

The meter can therefore be converted to a voltmeter reading 10 volts (FSD) by connecting 970 Ω in series with it.

It is therefore possible to make a multimeter to read a variety of currents and voltages by providing a bank of resistors which can be switched into the meter circuit either in series or parallel as required.

Use of ammeter and voltmeter

The moving coil meter is by far the most convenient way to measure current and voltage but it must be appreciated that it has to draw some current to operate and it must in consequence necessarily change the behaviour of the circuit under test.

Consider the two circuits in Fig. 26 to measure the value of an unknown resistance R by noting the current through it and the potential difference across it.

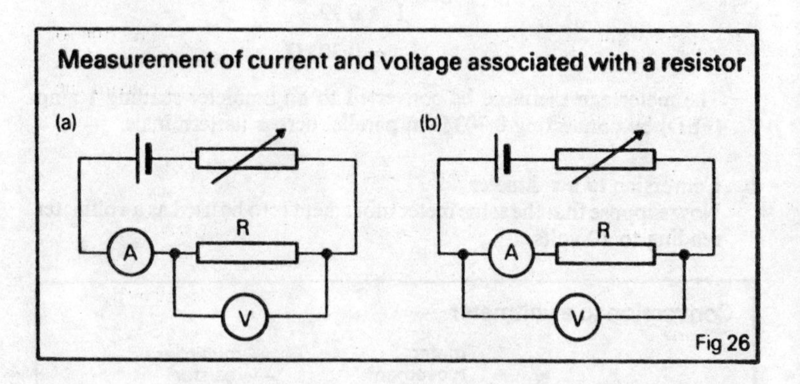

In Fig. 26a, which is the most common arrangement, the potential difference across R is being measured correctly but the measured current which is assumed to pass through R is in fact split between R and the voltmeter. This problem is not serious if the resistor value is much less than the voltmeter resistance since nearly all of the current will pass through R but in measuring high resistances the voltmeter will take an increasingly large share of the current and the errors will rise accordingly.

The problem of measuring the current through the resistor is solved in Fig. 26b but only at the expense of measuring the potential difference across both R and the ammeter. Now the error is acceptable provided R is large compared to the ammeter resistance when most of the potential difference will be across R. Fig. 26a should therefore be used for low resistance circuits but Fig. 26b is preferred in the high resistance case.

The requirement that the moving coil meter draws a current to operate also presents difficulties in reading the e.m.f. of a cell, since connecting the meter across the cell sets up a voltage drop across the internal resistance of the cell (the so called 'lost volts'). This voltage drop is only negligible if the voltmeter resistance is very high compared to the internal resistance of the cell and for accurate work a **null method** is preferred using a **potentiometer** as discussed in the next chapter.

Worked examples on chapter 2

1 a Calculate the magnetic flux density at a distance of 2 cm from a long straight vertical wire which is carrying a current of 20 A. ($\mu_0 = 4\pi \times 10^{-7}\ \text{Nm}^{-1}$)

b If the horizontal component of the earth's magnetic field is 2×10^{-5} T, calculate the distance from the wire at which the resultant flux will be zero.

a For a long straight wire the magnetic flux density is given by:

$$B = \frac{\mu_0 I}{2\pi r} = \frac{4\pi \times 10^{-7} \times 20}{2\pi \times 2 \times 10^{-2}} = 20 \times 10^{-5} = 2 \times 10^{-4}\ \text{T}$$

b If the resultant flux density is to be zero then the flux density due to the wire will need to be 2×10^{-5} T so that cancellation can occur.

Since flux density is inversely proportional to distance from the wire we can write:

$$B \propto \frac{1}{r} \quad \text{or} \quad B = \frac{K}{r} \quad \text{where K is a constant}$$

When $B = 2 \times 10^{-4}$ T, $r = 2$ cm (2×10^{-2} m) hence:

$$K = Br = 2 \times 10^{-4} \times 2 \times 10^{-2} = 4 \times 10^{-6}$$

Hence when B is equal to 2×10^{-5} T, distance r is given by:

$$r = \frac{K}{B} = \frac{4 \times 10^{-6}}{2 \times 10^{-5}} = 2 \times 10^{-1} = 20\ \text{cm}$$

2 Two moving coil galvanometers X and Y are identical except for the following difference in specification:

Coil X – resistance 30 ohms 100 turns
Coil Y – resistance 80 ohms 200 turns

Compare **a** the current sensitivity, **b** the voltage sensitivity, **c** the deflection obtained when connected across an e.m.f. of internal resistance 50 ohms.

a Current sensitivity $= \frac{\theta}{I} = \frac{BAn}{K} = Cn$ where C is a constant

$$\therefore \frac{\text{sensitivity X}}{\text{sensitivity Y}} = \frac{C \times 100}{C \times 200} = \frac{1}{2}$$

b Voltage sensitivity $= \frac{\theta}{V} = \frac{BAn}{KR} = \frac{Cn}{R}$ where C is a constant

$$\therefore \frac{\text{sensitivity X}}{\text{sensitivity Y}} = \frac{C \times 100/30}{C \times 200/80} = \frac{100}{200} \times \frac{80}{30} = \frac{4}{3}$$

c Since deflection $\propto nI$ and $I \propto \dfrac{1}{\text{total circuit resistance}}$

$$\therefore \text{deflection} \propto \frac{n}{R_{total}} = \frac{Kn}{R_{total}} \quad \text{where K is a constant}$$

$$\therefore \frac{\text{deflection X}}{\text{deflection Y}} = \frac{K\,100}{K\,200}\,\frac{(50+80)}{50+30} = \frac{100}{200} \times \frac{130}{80}$$

$$\text{Hence ratio of deflections} = \frac{13}{16}$$

3 A battery of negligible internal resistance is used to drive a current through two resistors of 5000 ohms and 3000 ohms joined in series. A voltmeter of resistance 2000 ohms reads 6 volts when placed across the 3000 ohms resistor.

a What is the e.m.f. of the battery?

b What would the voltmeter read if placed across the 5000 ohm resistor?

The important point in this problem is that the resistance of the voltmeter is of the same order of magnitude as the resistors in the circuit and the voltmeter will therefore be drawing a significant part of the current.

a The total circuit resistance is $5000 + \left(\dfrac{3000 \times 2000}{3000 + 2000}\right)$

$$\left[\frac{\text{product}}{\text{sum}} \text{ for two resistors in parallel}\right]$$

$$\therefore R_{total} = 5000 + 3000 \times \frac{2}{5} = 5000 + 1200 = 6200 \text{ ohms}$$

$$\therefore I = \frac{V}{R} = \frac{6}{1200} \text{ amp}$$

(6 volts appear across the parallel combination of equivalent resistance 1200 ohms)

The battery of e.m.f. E volts is therefore driving 6/1200 amp through a single equivalent resistor of 6200 ohms and the value of E is therefore given by:

$$E = IR = \frac{6}{1200} \times 6200 = 31 \text{ volts}$$

3 b The parallel resistance is now:

$$\frac{5000 \times 2000}{5000 + 2000} = 500 \times \frac{2}{7} = 1428 \text{ ohms}$$

and the single equivalent resistor is:

$$(1428 + 3000) = 4428 \text{ ohms}$$

The 31 volt battery is therefore driving a current of $I = 31/4428$ amp through the parallel combination (resistor and voltmeter) of 1428 ohms.

The voltmeter reading is:

$$V = IR = \frac{31}{4428} \times 1428 = 10 \text{ volts}$$

Practice questions

1 A moving coil meter movement has a full scale deflection (FSD) when a current of 1 mA is flowing through it. If its resistance is 100 ohms, how would you convert it to read:
 a a current up to 1A
 b a potential difference up to 10 volts?

2 A horizontal straight wire is 6 cm long and has a mass of $1.0\ \text{gm}^{-1}$ and a resistance of $4\,\Omega\text{m}^{-1}$. Calculate the potential difference that must be applied across the ends of the wire if it is to support itself in a uniform horizontal magnetic field of flux density 0.5 T. ($g = 9.8\ \text{m s}^{-5}$.)

3 Two long straight parallel wires 10 cm apart carry a current of 10 A and 5A respectively in the same direction. Where must a third current carrying conductor be placed so that it will experience no force?

4 A solenoid is 50 cm long and possesses 2000 turns of wire. When a current of 2 amps is passed through the coil the magnetic flux density at the central iron core is 1 T. What is the permeability of the iron under these conditions?

3 THE POTENTIOMETER AND WHEATSTONE BRIDGE

The principle of the potential divider

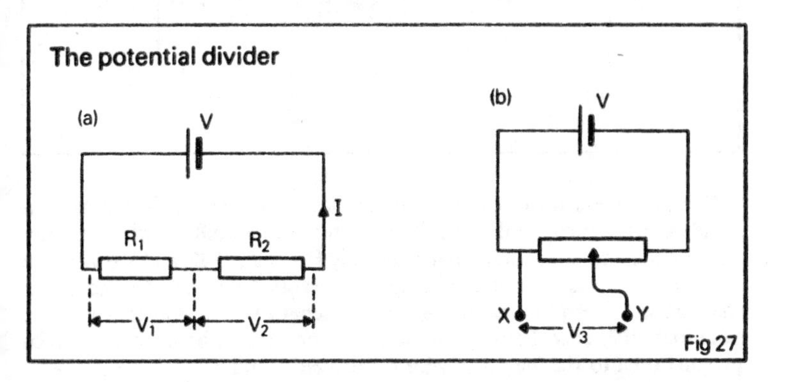

Fig 27

Consider Fig. 27a where the current (I) in the circuit is given by:

$$I = \frac{V}{R_1 + R_2}$$

The potential difference V_1 across R_1 is therefore given by:

$$V_1 = IR_1 = \frac{VR_1}{R_1 + R_2}$$

and similarly:

$$V_2 = \frac{VR_2}{R_1 + R_2}$$

Dividing V_1 by V_2 gives $V_1/V_2 = R_1/R_2$ and the total potential difference V is therefore divided in the ratio of the fixed resistors R_1R_2.

This idea can be used as in Fig. 27b to provide a variable voltage V_3 by connecting the slider of a rheostat as shown. If the terminals X and Y are connected to another circuit the input voltage to that circuit can be varied from zero to V volts ($0 \leqslant V_3 \leqslant V$) by moving the slider. A very effective means of voltage regulation is therefore available.

The potentiometer

The idea of resistors in series dividing the available potential difference is used in the slide wire potentiometer shown in Fig. 28.

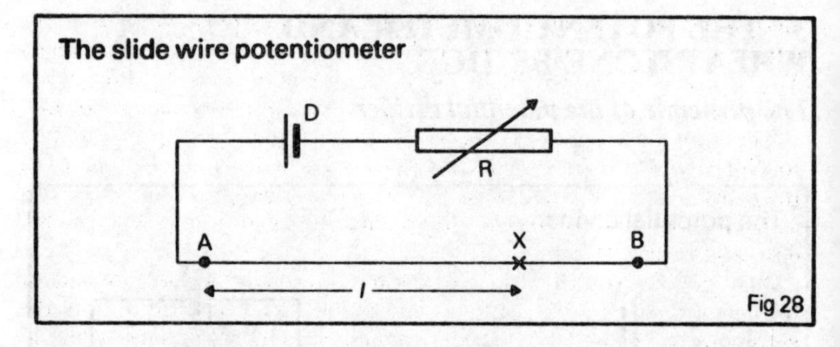

D represents a driver cell (usually an accumulator) which maintains a steady current through the circuit. AB consists of a uniform wire typically about one metre long, of resistance usually less than 10 Ω, and the variable resistor R enables the potential difference across AB to be adjusted. Since the wire is uniform its resistance per unit length will be constant and the potential difference between A and any point on the wire X will be proportional to the length of wire ℓ between A and X.

$$\therefore \; V_{AX} \propto \ell$$

$$\therefore \; V_{AX} = K\ell \quad \text{where K is a constant}$$

Now consider connecting into the circuit a test cell of e.m.f. E volts.

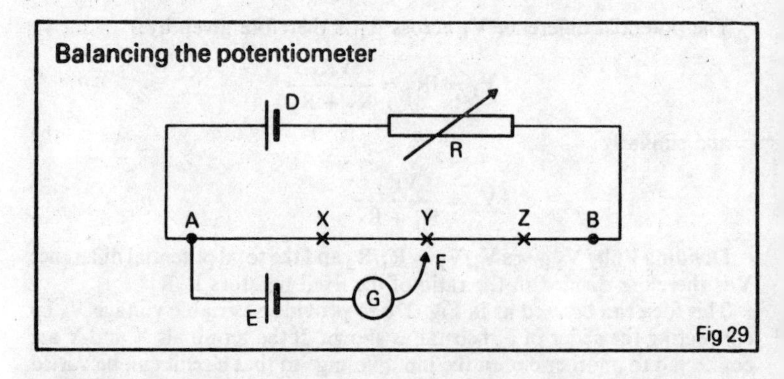

This circuit can be conveniently split into two parts consisting of the driver circuit as discussed previously and the test circuit which contains the cell E. The point F represents a 'flying lead' which can be touched onto the wire at any point along its length whilst G is a galvanometer so that any current flowing in the test circuit can be deflected.

Suppose that $V_{AX} < E$ whilst $V_{AZ} > E$ and consider the result of connecting F to X and Z in turn. In each case the centre zero galvanometer will register a deflection but the direction will be different in the two cases since with F at X the current is in the direction of E whilst with F at Z the current is in opposition to E. There should, however, be one point Y where $V_{AY} = E$ and the galvanometer shows no deflection since V_{AY} and E are acting in opposite directions and are therefore cancelling each other out.

The point Y can be found using the method of trial and error by carefully touching F onto the wire at different points until no deflection is observed in the galvanometer. It is important not to slide F along the wire since this tends to spoil the uniformity destroying the relationship $V_{AX} = K\ell$ discussed earlier.

N.B. Null method – Since no current flows through the galvanometer at the balance position, the resistance of the cell E and the resistance of the galvanometer will not alter the balance point. When the test cell is balanced at Y, the full e.m.f. is being balanced by V_{AY} and this null method of voltage measurement is not subject to the problem of 'lost volts'.

The potentiometer therefore possesses the advantage of accurate voltage measurement but the disadvantage of being more difficult to use than the ordinary moving coil meter.

Failure to balance the potentiometer

If after setting up the circuit of Fig. 29 it is found that a balance point cannot be achieved, the most likely reasons for failure to balance are:

a Test cell E connected the wrong way round. The driver and test cells need to be connected with like poles joined to the same end of the wire – reverse the test cell connections if necessary.

b Test cell e.m.f. greater than the potential difference across the wire. This means that there will be no point on the wire capable of balancing the cell. This condition could be caused by:

   i Test cell e.m.f. > driver cell e.m.f. – increase the driver e.m.f. by adding more cells in series.

   ii The resistance of the rheostat R is too high. This means that although the driver cell possesses a sufficiently high e.m.f., too many volts are appearing across the rheostat leaving too small a potential difference across the wire itself – move the slider of the rheostat to one end such that the resistance is zero and try to balance the potentiometer once again. (The rheostat could, of course, be eliminated completely from the circuit but it is required so that the balance length, once achieved, can be adjusted.)

   iii High resistance connections in the driver circuit – make sure that all of the connections are tightened down correctly and if necessary clean the contacts with emery paper or some other suitable light abrasive.

c Broken connections or a faulty galvanometer.

N.B. After setting up the circuit, considerable time can be saved by connecting the flying lead to each end of the wire in turn. If opposite galvanometer deflections are not obtained then the system will not balance at any point along the length of the wire and the procedure described above should be carried out immediately.

Experiments with a potentiometer

1 Measurement of an unknown e.m.f. (E)

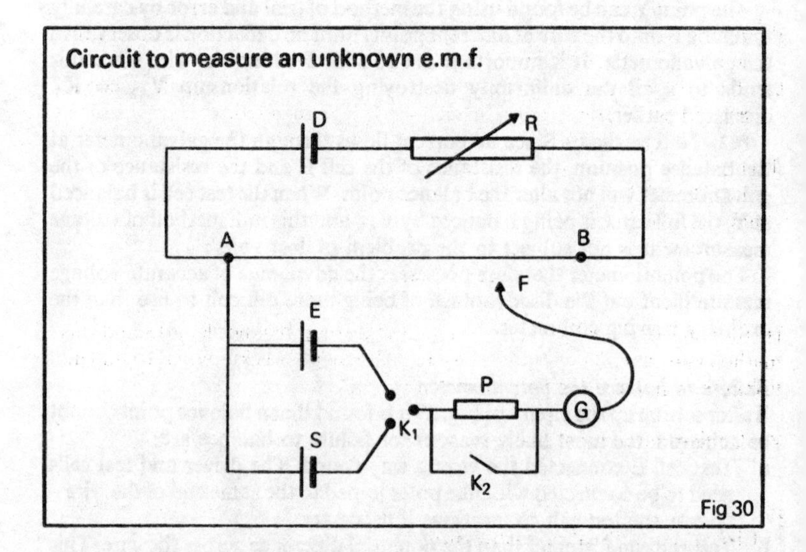

Fig 30

S represents a standard cell whose e.m.f. is known accurately. A typical example is the Weston cadmium cell which is **not** suitable in the driver circuit as a source of current since the e.m.f. changes appreciably if a current in excess of about 10 microampères is drawn from it. The purpose of the driver cell **D** is to send a steady current through the potentiometer wire and in general standard cells are not suitable for this function since they can neither deliver a large current nor produce a steady current for a long period of time.

K_1 is a key switch such that either E or S can be separately switched into the circuit.

P is a protective resistor used partly to protect the galvanometer from large deflections which might occur if F is placed a long way from the balance point and partly to protect the standard cell S. Since this is a null method, the presence of P will not change the balance position but once balance is approximately achieved, P is usually shorted out using K_2 so that an accurate balance can be determined. When K_2 is closed care must be taken to ensure that F is only touched onto the wire in close proximity to the balance position otherwise unwanted large deflections may result.

Experimental procedure

a Balance E and S separately.

b Adjust R so that the larger of the two balance lengths is almost the full length of the wire.
c Measure ℓ_1 (the balance length of E) and ℓ_2 (the balance length of S) from the end A of the wire (the end to which the cells are attached). R is used to give long lengths and consequently greater accuracy.
d Adjust R to obtain a series of value for ℓ_1 and ℓ_2 making sure that the lengths are always at least half of the length of the wire.
e Turn the wire round and balance from the other end to obtain a second set of results. Average the results to reduce the error.

Theory

Since $E = K\ell_1$ where K is a constant
and $S = K\ell_2$

$$\therefore \frac{E}{S} = \frac{\ell_1}{\ell_2} \quad \text{so} \quad E = S\frac{\ell_1}{\ell_2}$$

i.e. the e.m.f. of the cells is in the same ratio as their balance lengths and this method can be used to compare two e.m.f.'s if neither is known or to find an unknown e.m.f. if a standard cell is available.

2 Calibration of a voltmeter

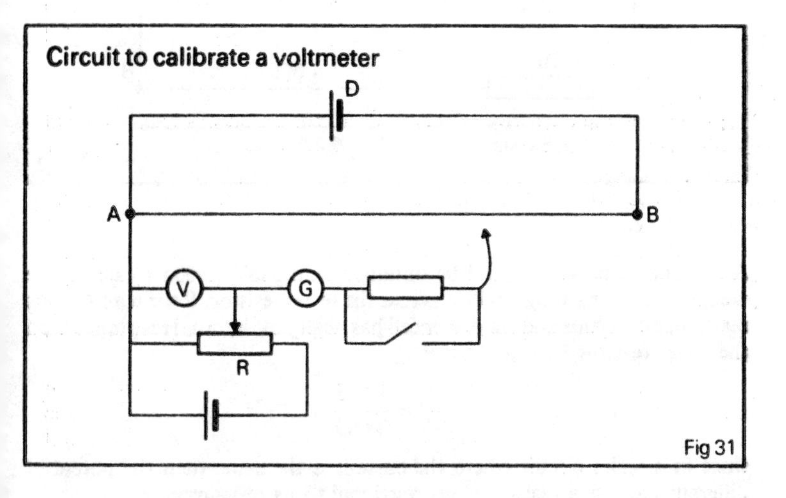

Fig 31

The following steps are required:
a The potentiometer wire (AB) is calibrated by first balancing a standard cell across it as discussed in the previous experiment. The potential drop per centimetre of wire can now be found and this enables any other balance to be converted into a potential difference.

b The lower circuit in Fig. 31 of cell and rheostat (R) provides an adjustable potential difference which is measured by both the voltmeter under test and the calibrated potentiometer wire.

c A number of readings are taken by adjusting the rheostat to check the calibration and ensure that the current through the potentiometer wire has not fallen. This might tend to happen if the driver cell (D) was running down.

3 Measurement of a small e.m.f.

Consider attempting to measure a thermocouple e.m.f. of about 4 millivolts using this potentiometer method. Since a typical driver cell e.m.f. is 2 volts, the balance point will be very close to one end of the wire. A series resistor (R_s) is therefore required so that the potential difference across the potentiometer wire is reduced.

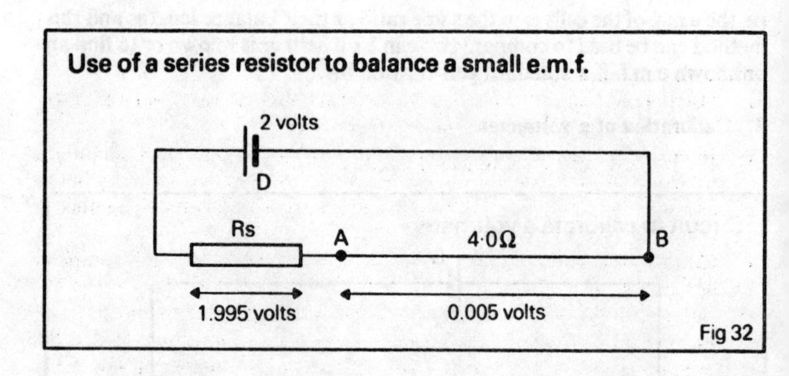

Fig 32

A reasonable balance would be obtained with 5 mV (0·005 V) across the wire and the remaining 1·995 V across the series resistor. If the wire AB has resistance 4·0 ohms and the driver cell has negligible internal resistance then the series resistor R_s is given by:

$$\frac{R_s}{4} = \frac{1{\cdot}995}{0{\cdot}005}$$

since in a series circuit where the current is the same then the potential difference across a resistor is proportional to its resistance.

$$\therefore R_s = 1596\,\Omega$$

This calculation gives a rough idea of the series resistor which will be required to balance the thermocouple but in practice a standard cell is needed since assuming a driver cell potential difference of 2 volts is not sufficiently accurate.

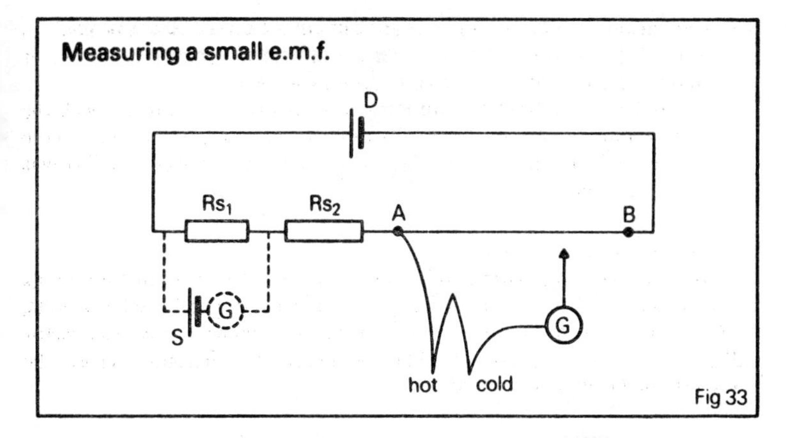

Fig 33

Experimental procedure

a Use two resistance boxes (Rs_1 and Rs_2) arranged in series such that $Rs_1 + Rs_2 = 1596\,\Omega$ (as calculated previously).

b Balance the thermocouple and adjust the series resistance slightly if necessary to produce a long balance length for greater accuracy (let us assume that no adjustment is required and that a reasonable balance at the 80 cm mark is achieved).

c Connect the standard cell (s) [e.m.f. = 1·018 volt] and galvanometer across one resistor only (dotted lines on Fig. 33).

d Balance the system (zero galvanometer reading) by adjusting Rs_1 and Rs_2 such that ($Rs_1 + Rs_2$) remains equal to 1596 ohms. Hence the potential difference across Rs_1 will change but the current in the driver circuit will remain constant and the potential difference across the wire (AB) will not alter.

e Note the value of Rs_1 to balance the standard cell S and hence find the potential difference per ohm in the circuit.

For example suppose $Rs_1 = 800$ ohms (and therefore $Rs_2 = 796$ ohms).

Since the galvanometer shows zero deflection, Rs_1 has 1·018 volt across it.

$$\therefore \text{ Potential difference per ohm} = \frac{1{\cdot}018}{800}$$

$$= 0{\cdot}00127 \text{ volt ohm}^{-1}$$

f The system is now calibrated and the standard cell can be removed. The galvanometer is re-connected to the thermocouple and the balance length determined accurately.

Calculation of the e.m.f.:

Suppose that the balance is at the 80 cm mark on a 100 cm wire.
At 0·00127 volt ohm^{-1} the potential difference across a 4·0 Ω wire is 4 × 0·00127 volt
∴ p.d. across 80 cm of wire is:

$$4 \times 0{\cdot}00127 \times \frac{80}{100} = 4{\cdot}06 \times 10^{-3} \text{ volts}$$

4 Measurement of the internal resistance of a cell

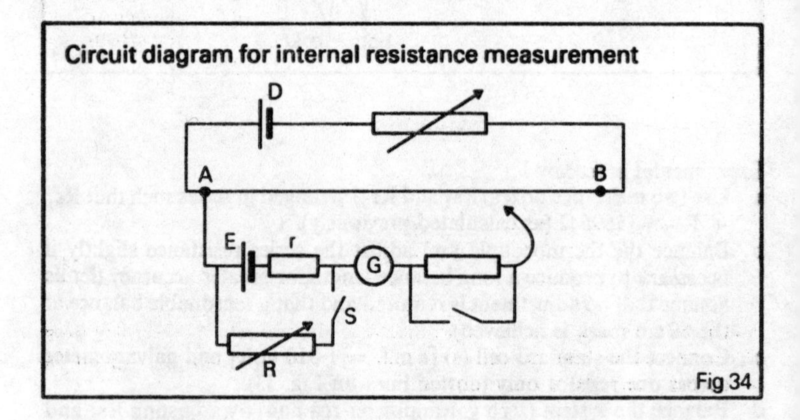

Fig 34

The test cell E with internal resistance r ohms is first balanced with S open at a distance ℓ_1 from the end A to give a measure of the e.m.f. of the cell. When S is closed a current flows through the standard variable resistor R. Some volts are now 'lost' in the battery itself and the potential difference across its terminal falls to V volts and the balance length is, therefore, reduced to ℓ_2.

Since $$E = I(R + r)$$
and $$V = IR$$
when I is the current through R

$$\therefore \frac{V}{E} = \frac{IR}{I(R + r)}$$

$$= \frac{R}{R + r} = \frac{\ell_2}{\ell_1}$$

Rearranging gives $r = R\left(\frac{\ell_1}{\ell_2} - 1\right)$

A number of readings can be taken and averaged by varying and noting the corresponding value of ℓ_2.

5 Comparison of resistances

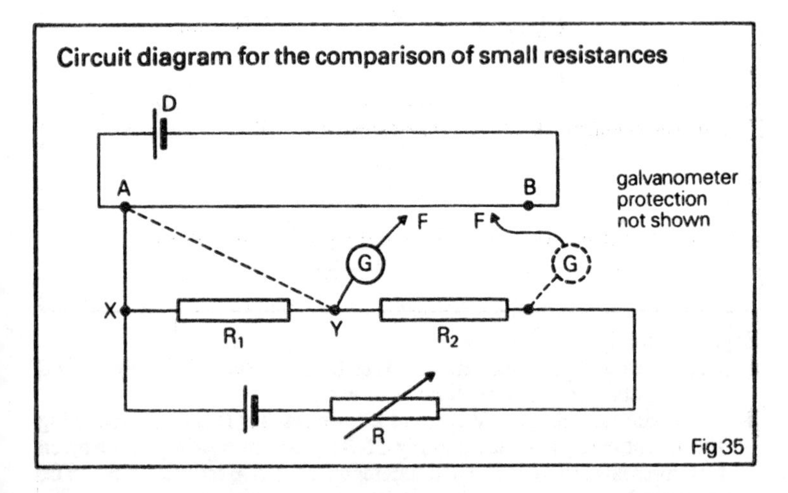

Fig 35

When the resistors R_1R_2 to be compared are arranged in series, they pass the same current and the potential differences across them (V_1V_2) can be compared using the potentiometer.

Since $$V_1 = IR_1$$

and $$V_2 = IR_2$$

Hence $$\frac{V_1}{V_2} = \frac{R_1}{R_2} = \frac{\ell_1}{\ell_2}$$

The resistors can therefore be compared by comparison of their balance lengths and a series of readings can be taken by adjusting the rheostat R.

N.B. i This method is especially useful for the comparison of **small resistances** since the resistance of the connecting wires will not affect the result.

ii The resistors being compared need to be of similar value otherwise a long balance length for one resistor will correspond to a short length for the other and the experimental error will increase.

6 Measurement of current and calibration of an ammeter

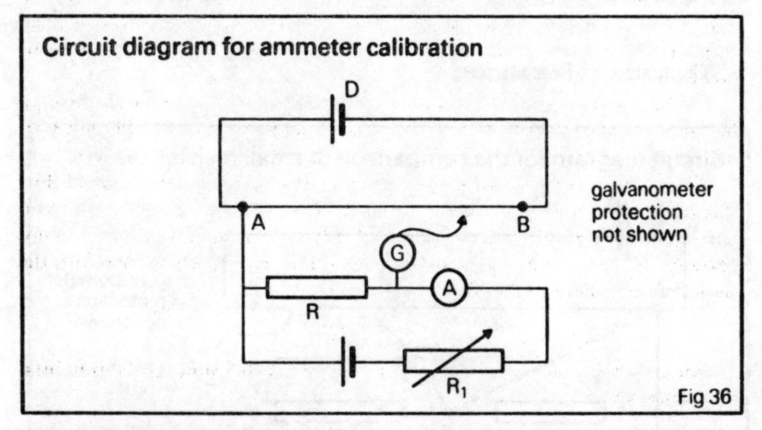

Fig 36

Experimental procedure

a Calibrate the wire using a standard cell (as discussed previously) to find the potential drop per centimetre of wire.

b The potential difference V across a known resistor R is measured using the potentiometer as shown in Fig. 36. The current flowing is then given by Ohm's law as $I = V/R$ and ammeter connected in series with R can be calibrated.

c A number of readings are taken by varying the rheostat R_1 so that the ammeter can be checked over all of its scale.

The Wheatstone bridge

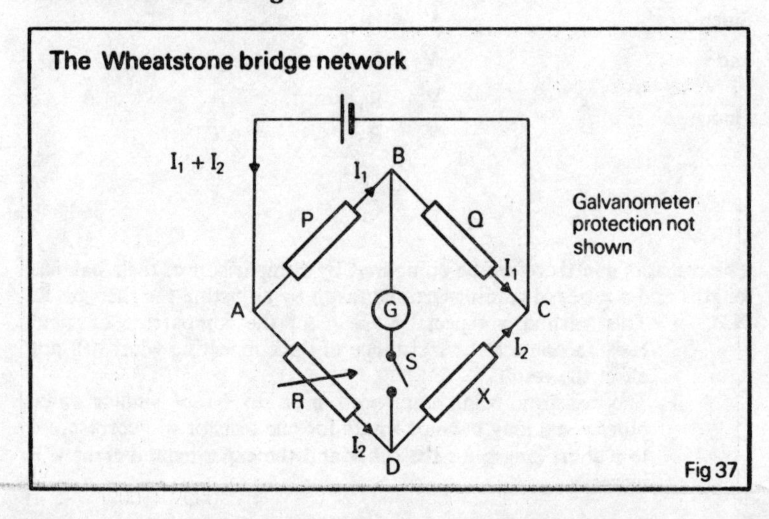

Fig 37

Consider the 'bridge' of four resistors shown in Fig. 37 where a centre zero galvanometer is connected across one pair of opposite corners of the square and a cell is connected across the other pair. P and Q are known fixed resistors, R is known but variable and X is the unknown which is to be determined.

The bridge is balanced by adjusting R such that no current flows through the galvanometer when the tapping key (S) is depressed (i.e. the galvanometer shows no deflection).

When this balance condition is achieved, the same current I_1 must flow through both P and Q whilst a current I_2 flows through R and X. Since no current flows through the galvanometer, there can be no potential difference between B and D and the potential difference across AB is therefore the same as the potential difference across AD

$$\therefore\ I_1P = I_2R$$

Similarly the potential difference across BC is the same as the potential difference across DC and so:

$$I_1Q = I_2X$$

Dividing one equation by the other eliminates the two currents and produces the final expression:

$$\frac{P}{Q} = \frac{R}{X}$$

This simple expression involving only the four resistances allows the unknown X to be found provided the other three resistors are known and the accuracy of the method is limited only by the tolerance on the known resistors provided the galvanometer is reasonably sensitive.

Practical Wheatstone bridge arrangements

1 The metre bridge

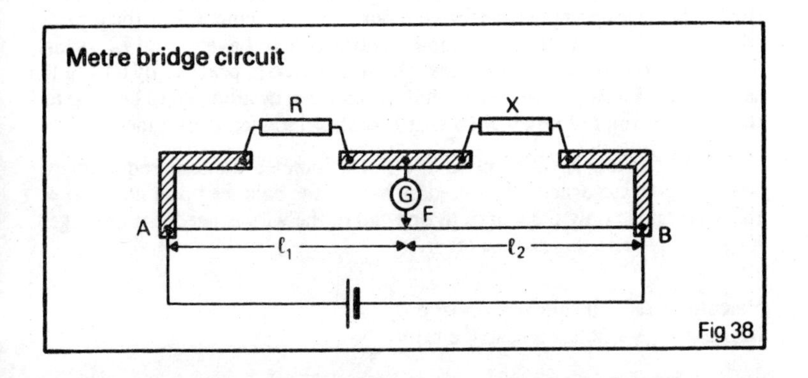

Fig 38

In this arrangement R is a standard resistance and X is the unknown whose value is to be determined. The shaded part of Fig. 38 represents connections made with strips of copper or brass so that the resistance is negligible. The other two resistors in the bridge network as discussed previously are provided by the resistance wire AB of resistance ρ ohms per metre. This wire can be split into two resistors AF and FB of value $\rho\ell_1$ and $\rho\ell_2$ by placing the flying lead F a distance ℓ_1 from A and ℓ_2 from B. The ratio of these two resistors can be quickly altered simply by moving F to another part of the wire and balancing the arrangement is therefore easier than for the type of bridge discussed in the previous section.

When the bridge is balanced (zero galvanometer deflection when F makes contact with AB), the equation linking the four resistors can be applied and we can write:

$$\frac{R}{X} = \frac{\rho\ell_1}{\rho\ell_2} = \frac{\ell_1}{\ell_2}$$

Provided R is known then X can be found simply by measuring the lengths from the flying lead to each end of the wire (or one length will be sufficient if the wire is known to be one metre long as is often the case when $\ell_1 + \ell_2 = 1$).

Comparison between metre bridge and potentiometer

Whilst there are some important points of similarity between the two circuits it is essential to realise that the **balancing requirements are different.**

Since the lengths ℓ_1 and ℓ_2 are both required it is necessary to balance the bridge near the centre of the wire for greatest accuracy since both lengths need to be as long as possible for an accurate measurement. For the typical wire of length one metre it is a good working rule to have ℓ_1 and ℓ_2 within the range forty to sixty centimetres. This is achieved in practice by having R as a variable known resistance so that its value can be adjusted to be close to the value of the unknown X so that a central balance is obtained.

This condition is different to the potentiometer balance requirement where greater accuracy is favoured by having the balance point near one of the wires, since only the length to one end of the wire is needed and longer length measurements reduce error.

Potentiometer – balance near one end
Metre bridge – balance near the centre

2 The post office box

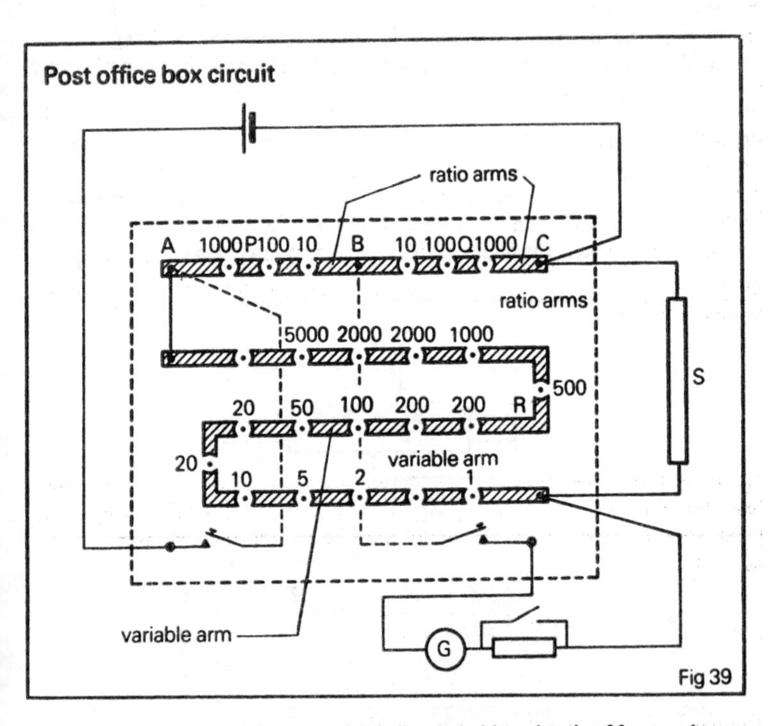

Fig 39

Close inspection reveals once again the basic bridge circuit of four resistors connected by galvanometer and cell. In this arrangement, three of the resistors PQR consist of plug resistance boxes permanently connected together whilst the unknown resistance X and the galvanometer and cell are connected externally when required. P and Q can only take the values 10, 100 or 1000 ohms whilst R is adjustable in steps on one ohm up to, typically, 10,000 Ω.

P and Q are known as the ratio arms of the bridge and the ratio P : Q can be varied from 100 : 1 to 0·01 : 1 by removing the appropriate plugs. A wide range of resistance values for X can be measured accurately by varying R and balancing the bridge. The apparatus can be used quickly and easily once the operator is familiar with the balancing technique but the experimental procedure for determining an unknown resistance will not be discussed here.

Limitations of the Wheatstone bridge

a Resistor tolerance.

b Contact resistance at the connections between different components. This problem can be reduced in the metre bridge by interchanging the standard and unknown resistors and repeating the experiment to average the results.

c When the **unknown resistance is small** (one ohm or less) contact resistance errors are likely to be appreciable and the **potentiometer** method discussed earlier is **more suitable.**

Worked examples on chapter 3

1 A load resistor of value 200 Ω is connected to a 20 volt supply via a potential divider of resistance 600 Ω so that the potential difference across the load can be varied. What is the value of this potential difference when the slider is in the half way position?

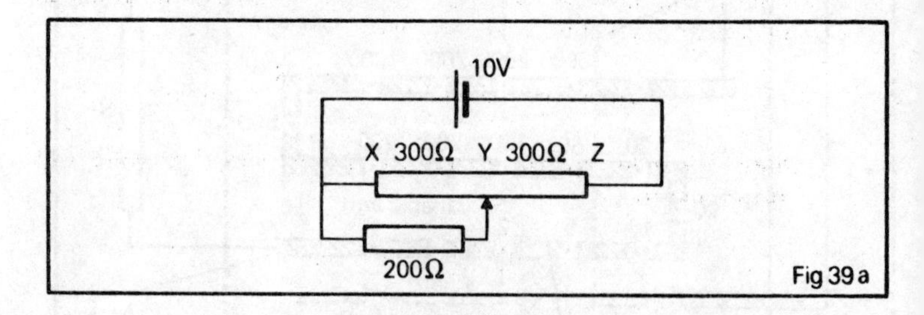

Fig 39 a

$$R_{xy} = \frac{300 \times 200}{300 + 200} = 300 \times \frac{2}{5} = 120\,\Omega$$

$$\therefore\ R_{xz} = 120 + 300 = 420\,\Omega$$

Since the potential is being divided in the ratio of the resistors:

$$V_{xy} = 10 \times \frac{120}{420} = 2{\cdot}86 \text{ volts}$$

∴ The potential difference across the load resistor is 2·86 volts.

2 The circuit diagram below shows a potentiometer wire of length one metre and resistance 4 ohms with a driver cell of e.m.f. 2 volts and negligible internal resistance. R is a variable resistor in the driver and E is the test cell which is to be balanced.

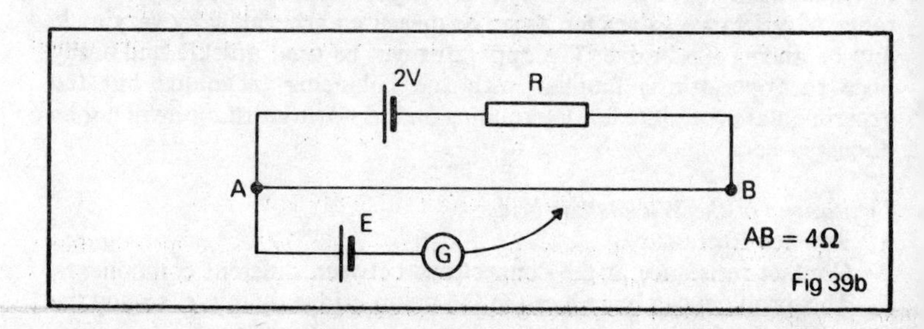

Fig 39b

a What is the balance length with E = 1·5 volt and R = 1 ohm?

b What is the new balance length if the internal resistance of E is 1 ohm and a 20 ohm resistor is inserted in series with E and the galvanometer G?

c What is the new balance length if the 2 ohm resistor is now placed across E?

d Referring to the original diagram, explain why the system will not balance if R = 4 ohms.

e Referring to the original diagram, if R = 20 ohms and the balance length is 90 cm, what is the e.m.f. of the test cell.

2 a If R = 1 Ω and the wire has resistance 4 Ω then the total series resistance in the driver circuit is 5 Ω and four fifths of the available potential drop will be across the wire

$$\therefore \text{ p.d. across the wire} = 2 \times \frac{4}{5} = \frac{8}{5} = 1{\cdot}6 \text{ volt}$$

Since 1·6 volts appear across 100 cm of wire, 1·5 volts will correspond to:

$$100 \times \frac{1{\cdot}5}{1{\cdot}6} = 93{\cdot}8 \text{ cm}$$

$\therefore$ The balance length is 93·8 cm

b Introducing series resistance into the test circuit will not alter the balance length since no current is flowing through the test cell when the system is balanced. The only change is that the galvanometer deflections will be less when the cell is not balanced and the balance will be more difficult to detect accurately.

c When the resistor is placed in parallel across E the test cell will now drive a current through the resistor and its own internal resistance. The current flowing is:

$$I = \frac{V}{R} = \frac{1{\cdot}5}{3} = 0{\cdot}5 \text{ amp}$$

(total resistance = 3 Ω since the cell internal resistance must be included)

$\therefore$ The lost volts are equal to 0·5 volt (0.5 × 1) and the potential difference across the cell terminals therefore falls to 1 volt.

Since 1·6 volts appear across 100 cm of wire, 1 volt will correspond to:

$$100 \times \frac{1}{1{\cdot}6} = 62{\cdot}5 \text{ cm}$$

$\therefore$ The new balance length is 62·5 cm

d When R = 4 ohms the potential difference will be divided so that 1 volt appears across R and 1 volt appears across the wire AB. A 1·5 volt test cell will, therefore, not balance.

e If $R = 2\,\Omega$ then the total series resistance in the driver circuit is $6\,\Omega$ (2 + 4). The potential drop across the wire is therefore:

$$\text{p.d. across AB} = 2 \times \frac{4}{6} = 1{\cdot}33 \text{ volts}$$

Since 1·33 volts appear across 100 cm of the wire then the potential drop across 90 cm of wire is:

$$\text{p.d. across 90 cm} = 1{\cdot}33 \times \frac{90}{100} = 1{\cdot}2 \text{ volts}$$

$$\therefore \text{ e.m.f. of test cell} = 1{\cdot}2 \text{ volts}$$

3 **A metre bridge has a length of wire in the left hand gap and a resistor R in the other gap. At 0°C the balance is 40·0 cm from the left hand end and this increases to 43·0 cm when the temperature of the wire is increased to 100°C. What is the temperature coefficient of resistance of the material of the wire?**

Let the initial wire resistance be X_0.
When the bridge is balanced we can write:

$$\frac{X_0}{R} = \frac{40}{60} \quad \text{so} \quad X_0 = \frac{40\,R}{60} = 0{\cdot}67\,R$$

Similarly $$\frac{X_{100}}{R} = \frac{43}{57} \quad \text{so} \quad X_{100} = \frac{43\,R}{56} = 0{\cdot}75\,R$$

The temperature coefficient of resistance (α) is given by:

$$\alpha = \frac{X_{100} - X_0}{X_0 \times 100} = \frac{0{\cdot}75\,R - 0{\cdot}67\,R}{0{\cdot}67\,R \times 100}$$

$$= \frac{0{\cdot}08\,R}{67\,R}$$

$$= 1{\cdot}19 \times 10^{-3}\,K^{-1}$$

N.B. Care is needed with the accuracy of this result since the term ($X_{100} - X_0$) is a small quantity and any errors in X_{100} or X_0 can cause a large error when the difference is taken.

Practice questions

1 A slide wire potentiometer is to be used for the comparison of the e.m.f.'s of two cells. Discuss the experimental procedure which should be adopted and also comment on the following points:

a The main advantage of using a potentiometer for the comparison.
b The qualities needed in the driver cell.
c The reasons for incorporating a rheostat in the driver circuit.
d Possible reasons for failure to achieve a balance point along the wire.

Discuss how the circuit could be adapted so that an e.m.f. of a few millivolts produced by a thermocouple could be measured. Explain how the potentiometer wire is calibrated using a standard cell and suggest possible values for the components in the circuit if the wire itself has a resistance of 40 ohms.

2 Describe how you would compare two resistors using a metre bridge form of the Wheatstone bridge circuit. Derive from first principles the condition which is satisfied when the bridge is balanced.
Why is the method not satisfactory if the two resistors are either very small or of widely differing values?

3 A cell of e.m.f. 2 volts is balanced across 90 cm of a potentiometer wire of length 1 metre. When a 5 ohm resistor is placed across the cell the balance length falls to 80 cm. What is the internal resistance of the cell?

4 The left and right hand gaps of a metre bridge circuit contain resistors of value 3 ohms and 2 ohms respectively. When the 3 ohm resistor is shunted with wire the balance point moves to the 50 cm mark. If the shunt is one metre long and 0·2 mm diameter, calculate its resistivity.

4 ELECTROMAGNETIC INDUCTION

The discovery that an electric current had an associated magnetic field suggested that the converse might also be true and resulted in the discovery of the **dynamo effect**.

Dynamo effect The motion of a conductor in a magnetic field produces an e.m.f. across the ends of the conductor provided it is cutting the magnetic field lines as it moves.

Subsequent experiments showed that, in fact, it is **relative** motion which is important and it does not matter whether the conductor moves across the field lines or vice versa. It is, however, important that the relative motion is across the field lines and not parallel to them.

Laws of electromagnetic induction

1 **Faraday's law** The value of the induced e.m.f. is directly proportional to the rate at which the magnetic field lines are cut (i.e. the faster the motion the greater the induced e.m.f.).
2 **Lenz's law** The induced e.m.f. is in such a direction that it tends to oppose the change which caused it.

Lenz's law is, in fact, a necessary consequence of the concept of **conservation of energy**. Consider a magnet being inserted into a coil of wire.

Direction of the induced e.m.f.

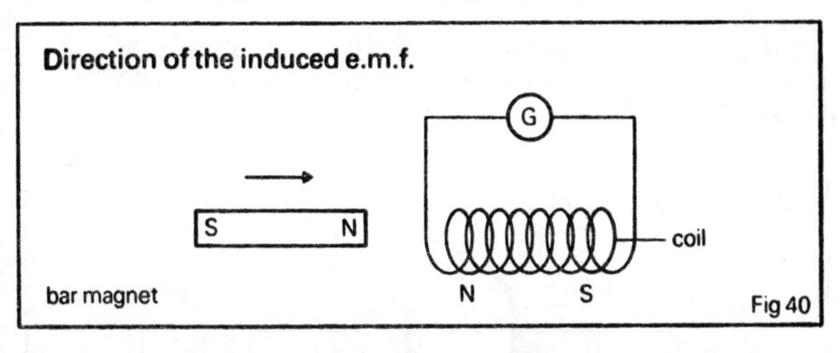

Fig 40

As the magnet enters the coil an e.m.f. is induced and a current flows which is registered on the galvanometer. Since the current has an associated magnetic field, the coil will act as a magnet with its poles arranged in such a way as to oppose the entry of the bar magnet. Work now has to be done in overcoming this force and it is this work which provides the electrical energy (current). The energy is finally dissipated as heat in the wire of the coil.

If the coil became magnetised in the opposite sense to that shown in Fig. 40 (in contravention of Lenz's law) then the velocity of the bar magnet would become greater. The system would then be producing an ever increasing amount of electrical energy without any work being done and this is clearly impossible.

Simple experiments show that the induced e.m.f. is greatest when the conductor is moved at right angles to both its own length and the magnetic field direction. The direction of the induced e.m.f. (and hence current where appropriate in a closed circuit) is given by Fleming's right hand rule.

Fleming's right hand rule

If the thumb, first finger and second finger of the right hand are held mutually perpendicular with the first finger in the magnetic field direction and thumb in the direction of motion then the direction of the second finger will indicate the direction of the induced e.m.f.

There is an obvious parallel here with Fleming's left hand rule discussed earlier:

Motor – left hand rule [current + magnetic field → motion]
Dynamo – right hand rule [motion + magnetic field → e.m.f.]

Magnetic flux

Unit Weber Wb symbol ϕ

This expresses the idea of a number of magnetic field lines passing through a particular area (A). The flux of magnetic field through a plane surface is the product of the area of the surface and the normal component of magnetic field (B). Hence:

$$\phi = BA$$

i.e. a field of strength B tesla will have BA lines of magnetic flux passing through an area A.

The flux is measured in Weber and since $B = \phi/A$ an alternative unit for magnetic field strength (flux density) is Weber per square metre ($Wb\,m^{-2}$) [$1T = 1\,Wb\,m^{-2}$]

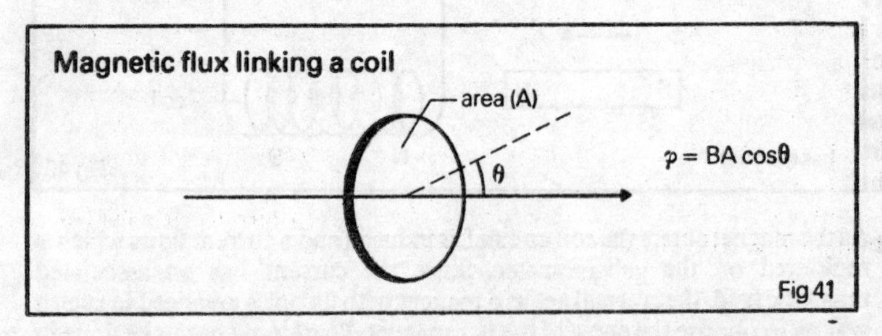

Fig 41

If the magnetic field direction is not at right angles to the coil (Fig. 41) the magnetic flux expression is modified to $\phi = BA\cos\theta$ and in the special case where $\theta = 90°$ the flux is zero since the field lines are no longer cutting through the area A.

Magnetic flux linkage

If the coil in Fig. 41 had N turns instead of one turn then the flux through the whole coil is the sum of the individual fluxes and is called the flux linkage given by $\phi = NBA\cos\theta$.

Flux linkage = number of turns × flux through the coil cross-section

Flux linkages are often expressed in **Weber-turns** to indicate clearly that the number of turns in the coil have been taken into account. This avoids possible confusion between flux linkage in a coil and flux through the coil cross-section.

Calculation of the induced e.m.f.

Consider a wire XY (Fig. 42) being propelled along a set of rails at constant speed V by a force F where the direction of motion is at right angles to a magnetic field of strength B (shown into the paper).

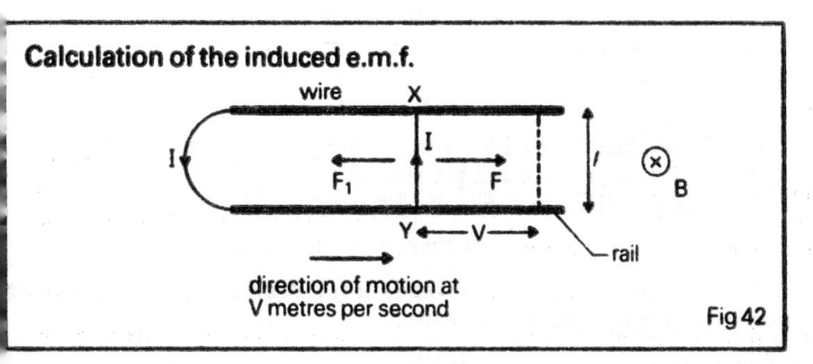

Fig 42

The dotted line a distance V away from the wire shows the position reached one second later. By Lenz's law an equal and opposite force (F_1) will be produced which will oppose the motion (thus satisfying Newton's law of motion that constant speed requires no resultant force).

[An alternative explanation is to say that XY will have a current induced into it and since we now have current in a conductor in a magnetic field there will be a force F_1 which will try to provide motion in the direction indicated. The use of Fleming's right hand rule to predict the current direction followed by Fleming's left hand rule to show the force will indicate that F_1 opposes F as shown on Fig. 42.]

Since the force on a conductor of length ℓ carrying a current I in a field of strength B is $BI\ell$ (chapter 2) we can write:

$$F_1 = F = BI\ell$$

Also

$$I = \frac{e}{r}$$

where e is the induced e.m.f. and r is the total resistance in the circuit

$$\therefore\ F = \frac{Be\ell}{r}$$

Since work done = force × distance moved

∴ Work done per second = force × distance moved per second

But the work done per second is the power and the distance moved per second is the velocity (V) and so:

$$\text{Power} = FV = \frac{Be\ell V}{r}$$

But

$$\text{power} = \frac{e^2}{r}$$

$$\left(\text{remember chapter 1: power} = \frac{(\text{potential difference})^2}{\text{resistance}}\right)$$

$$\therefore \frac{e^2}{r} = \frac{Be\ell V}{r}$$

$$\therefore e = B\ell V$$

But the product of ℓ and V is the area swept out by the wire per second.
∴ e = B × area swept out per second
Since $\phi = BA$ where ϕ is the flux we can write:

$$e = \text{rate of change of flux } (d\phi/dt)$$

Therefore the induced e.m.f. (e) and the rate of change of flux ($d\phi/dt$) are numerically equal and we can write:

$$e = -\frac{d\phi}{dt} \quad \textbf{Neumann's law}$$

The minus sign is a consequence of the induced current producing a field in opposition to the change of flux: Lenz's law.

This mathematical statement of the laws of electromagnetic induction is called Neumann's law.

Measurement of magnetic flux density using a search coil

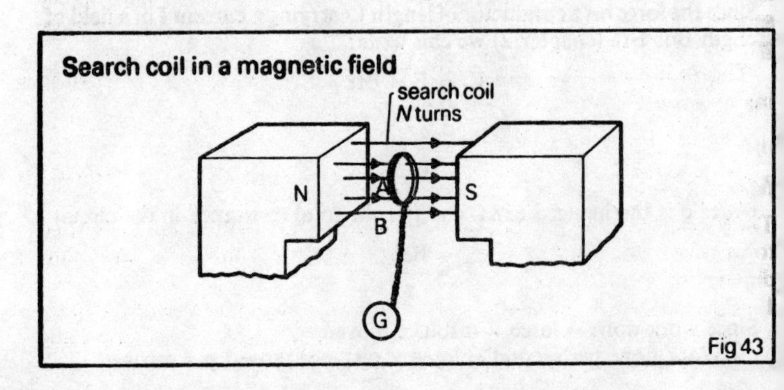

Fig 43

When the search coil is removed from the field the magnetic flux changes and a current is induced and measured on the galvanometer.

The induced e.m.f. is given by:

$$e = -d\phi/dt \quad \text{(Neumann's law)}$$

Hence the current flowing (I) is given by:

$$I = \frac{e}{r} = -\frac{1}{r}\frac{d\phi}{dt}$$

where r is the resistance in the circuit

Since current is the rate of flow of charge ($I = dQ/dt$) we can write:

$$\frac{dQ}{dt} = \frac{-1}{r}\frac{d\phi}{dt}$$

$$\therefore\ Q = \frac{-1}{r}\int_0^f d\phi = \frac{-1}{r}[\phi_t - \phi_0]$$

where ϕ_t and ϕ_0 are the fluxes linking the coil at times t and zero respectively.

$$\therefore\ Q = \frac{\text{change of flux linkage}}{r}$$

Here the minus sign has been omitted since we are only interested in the magnitude of the charge.

If the coil in Fig. 43 has a known area A and known number of turns N then the initial flux linkage will be $\phi = BAN$ whilst the coil is between the poles of the magnet. When the coil is removed the flux reduces to zero (ignoring the earth's magnetic field) and the flux change is therefore BAN.

$$\therefore\ Q = \frac{BAN}{r} \quad \text{and so} \quad B = \frac{Qr}{AN}$$

Since r can be measured B can therefore be found provided Q can be measured. This measurement of Q is best achieved by quickly removing the search coil from between the poles of the magnet and using a ballistic galvanometer (discussed in chapter 2) to measure the charge passed.

The calibration of the galvanometer could be accomplished by discharging a capacitor through it (discussed in chapter 6).

Measurement of earth's magnetic field

The same idea of moving a search coil across magnetic field lines can be used to measure the earth's magnetic field but there are two important differences:

1 Since the earth's field is uniform over a large area it is possible to use a large search coil (typically 30 centimetres square) but even then a large number of turns is required since the earth's field is so weak.

2 It is not now possible to remove the coil to a place of zero flux and a different technique is needed to obtain a change of flux. Consider the coil set horizontally so that the vertical component of the earth's magnetic field is cutting it normally. The flux through the coil is BAN where A is the area of the coil and N the number of turns. If the coil is turned through 180° the flux linkage will still be BAN but it now threads the coil in the opposite direction and the change of flux linkage is therefore, 2BAN. Rotating the coil through 180° will therefore cause a deflection on the ballistic galvanometer enabling the charge to be measured.

Since $Q = \dfrac{\text{change of flux linkage}}{r}$ where r is the resistance

Hence $$Q = \frac{2BAN}{r}$$

and $$B = \frac{Qr}{2AN}$$

The process could be repeated to find the horizontal component of the earth's field by starting with the coil set vertically and then rotating it through 180°.

e.m.f. generated by a rotating coil

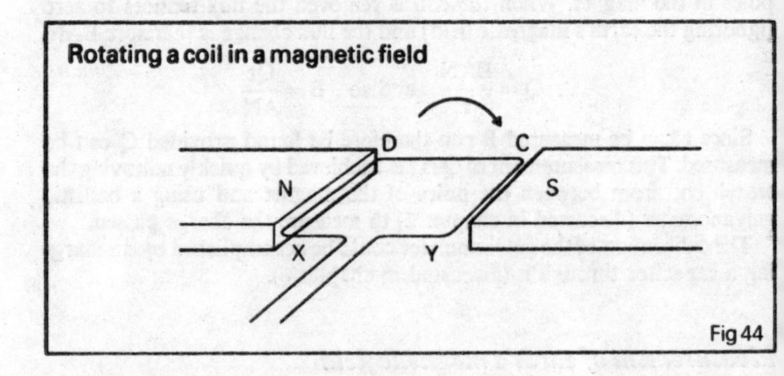

Consider a rectangular coil XYCD being placed in a magnetic field and rotated as shown. The sides XY and DC will not cut any flux and can therefore be ignored in terms of the generation of an e.m.f. whilst XD and YC will cut field lines but not at a constant rate. This is, perhaps, more easily seen by using a side view as shown in Fig. 45.

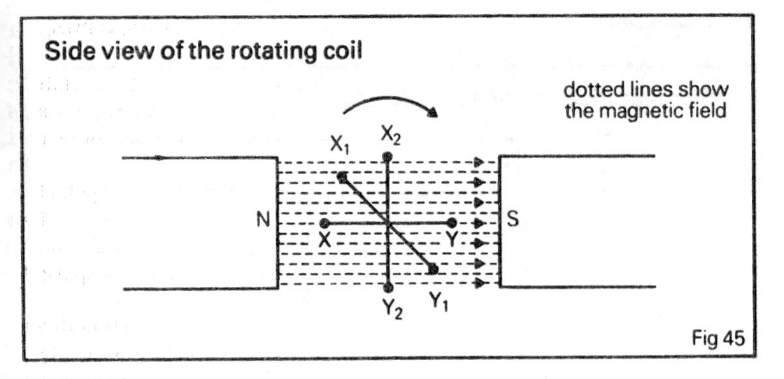

Fig 45

If the coil is rotating at a constant rate then the time taken to travel from XY to X_1Y_1 will be the same as the time to travel from X_1Y_1 to X_2Y_2 but the number of field lines cut will not be the same (i.e. the rate of cutting of field lines is not constant). The e.m.f. generated will, therefore, not be constant and will be a maximum when the coil is horizontal and zero when the coil is vertical.

Since $\phi = \text{NBA}\cos\theta$ (see Fig. 41)
and $e = -d\phi/dt$ (Neumann's law)

$$\therefore\ e = -\text{NBA}\frac{d}{dt}(\cos\theta) = \text{NBA}\sin\theta\, d\theta/dt$$

Since the angular velocity w is given by $W = d\theta/dt$ we can therefore write:

$e = \textbf{NBAW}\sin\theta$ showing that the output is sinusoidal

Commutators and slip rings are required for D.C. and A.C. generation respectively and have already been briefly discussed in the section on motors in chapter 2.

Eddy currents

When there is relative motion between a conductor and a magnetic field an e.m.f. is induced into all parts of the conductor. If the conductor is, for example, a large piece of metal then although the e.m.f. across the metal is likely to be low, the current flowing is likely to be large since the resistance is low.

These eddy currents oppose the motion (by Lenz's law) and mechanical energy is converted into electrical energy and then heat within the metal itself and the motion of the conductor is therefore damped down. Since this **eddy current loss** often needs to be **controlled** in iron cored apparatus such as motors, dynamos and transformers (to be discussed later) it is usual to build the iron parts with laminations. These are thin sheets of iron insulated from each other so that large currents (and hence energy losses) cannot build up.

The transformer

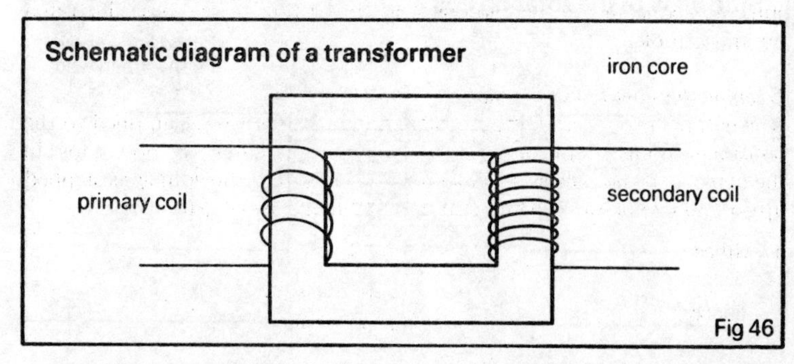

Fig 46

Consider two coils electrically unconnected but wound onto the same laminated iron core. An alternating current on one coil (the primary) will set up an alternating magnetic flux in the core and since this flux will cut the other coil (the secondary) an e.m.f. will be induced. The two coils are therefore linked together by magnetic flux and a change of current in either coil causes an e.m.f. to be induced in the other.

This type of electromagnetic induction is called **mutual induction** and forms the basis of a transformer. It can be shown that:

$$\frac{\text{secondary potential difference}}{\text{primary potential difference}} = \frac{\text{number of turns on the secondary}}{\text{number of turns on the primary}}$$

$$\text{i.e.}\ \frac{V_s}{V_p} = \frac{n_s}{n_p}$$

The potential difference can therefore be stepped up or down according to the turns ratio n_s/n_p. A **step-up transformer** has $n_s > n_p$ resulting in a greater potential difference in the secondary than the primary whilst a **step-down transformer** reduces the secondary output voltage.

If the transformer is one hundred per cent efficient (almost achievable in practice) there will be no power loss in the transfer of power between primary and secondary windings and we can write:

$$\text{power in primary} = \text{power in secondary}$$

$$V_s I_s = V_p I_p$$

where I_s and I_p are the currents in the secondary and primary coils respectively

Hence:

$$\frac{V_s}{V_p} = \frac{n_s}{n_p} = \frac{I_p}{I_s}$$

Increasing the turns ratio will therefore step up the potential difference but step down the current.

This ability to transform voltages is particularly useful in electrical power distribution since less energy is lost if the power is transmitted from power station to consumer at high voltage and low current. This is best illustrated by an example.

Example on power transmission

A generator produces 10 kw at 250 volts and the power is supplied to the consumer along a cable of resistance one ohm. Calculate the power loss in the cable, **a** If the power is transmitted directly, **b** If the voltage is stepped up to 2500 volts for transmission and stepped down again later.

a Since $P = VI$

hence $I = \frac{P}{V} = \frac{10000}{250}$

$= 40$ amps

$\therefore$ Power loss along the cable $= I^2R$

$= 40 \times 40 \times 1$

$= 1600$ watts

(i.e. approx 16% of the power is lost in the cable.)

b Transforming to 2500 volts reduces the current to 4 amps

$\therefore$ Power loss $= I^2R$

$= 4 \times 4 \times 1 = 16$ watts

The power loss is now reduced to about 0·16% provided the step up and step down transformers can be considered to be 100% efficient. Some power will, in fact, be lost in the transformers but the improvement is still significant.

The National Grid uses step up transformers so that power can be transmitted using voltages around 132,000 volts. This value is much too high for use in houses and factories and step down transformers are used to produce 240 volt mains voltage.

Self inductance

When a current flows through a coil of wire there is an associated magnetic field and if the current changes then the magnetic flux linking the coil will change and an e.m.f. will be induced. By Lenz's law this e.m.f. will oppose the change of current and will therefore act against the current if it is increasing but in the same direction if it is decreasing.

The growth and decay of the current is therefore slowed down in this process known as **self induction** and the induced e.m.f. is called a **back e.m.f.** since it opposes the change which caused it.

It is important to realise that this effect is only observable when the current is changing and inductive effects will therefore only be observed in steady D.C. circuits during switch on and off when the current is building to and decaying from its steady value.

Since the back e.m.f. (e) is proportional to the rate of change of current through the coil we can write:

$$e \propto dI/dt$$

$$\therefore\ e = -L\,dI/dt$$

where L is a constant known as the self inductance of the coil. The minus sign is incorporated since the back e.m.f. opposes the change of current which caused it and it is preferable to have the self inductance expressed as a positive constant.

Self inductance
Unit Henry Symbol L

Definition If the back e.m.f. is one volt when the current through the coil is changing at a rate of one ampere per second then the self inductance is one henry.

Since $e = -d\phi/dt$ (Neumann's law)
$e = -L\,dI/dt$ (definition of self induction)
Then the integration produces $\phi = LI$ and the self inductance can therefore be regarded as the flux linkage per unit current.

Mutual inductance
Unit Henry Symbol M

A similar concept can be applied for the so-called mutual inductance between two coils linked by magnetic flux in a transformer for example. Since the e.m.f. induced into the secondary coil depends on the rate of change of current in the primary we can write:

$$e \propto dI/dt$$

$$\therefore\ e = -M\,dI/dt$$

where the constant M is the coefficient of mutual inductance of the two coils.

Definition If an e.m.f. of one volt is induced into the secondary coil when the current changes at one ampère per second in the primary then the mutual inductance of the coils is one henry.

Non inductive winding
This phenomenon of self induction is a disadvantage in apparatus such as Wheatstone bridges and resistance boxes, since the resistors which need to be as accurate as possible are typically wound in coils of wire. If the bridge is not balanced then closing the galvanometer key will redistribute the current in the circuit such that some of it will now pass through the galvanometer which will deflect. These changing currents will, however, cause self-induction in the resistance coils and the achievement of a steady state reading on the galvanometer will be delayed. This can lead to errors in determining the balance point since near balance only a small deflection would be observed in any case and this could easily remain undetected when its build up is inhibited by self inductance in the resistance coils.

This problem can be reduced significantly by doubling the wire back on itself before coiling it up so that every part of the double coil has the current travelling in opposite directions resulting in a negligible magnetic field. This is known as winding the coil non inductively and is shown in Fig. 47.

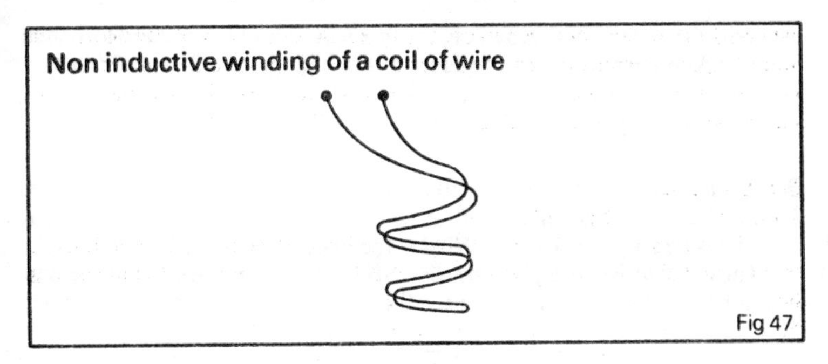

Fig 47

Energy stored in an inductor
Since a back e.m.f. is induced as the current increases in a coil of wire (an inductor) work must be done in increasing the current against this e.m.f. This energy supplied is stored in the inductor itself.
Rate of doing work = power = volts × amps

$$\therefore \text{ Rate of energy supplied} = L\frac{dI}{dt}.I$$

$$\therefore \text{ Energy supplied in a time } \delta t \text{ is } L\frac{dI}{dt}.I\delta t$$

$$\therefore \text{ Total energy supplied} = \int_0^I LI\,dI = \frac{1}{2}LI^2$$

$$\therefore\ \therefore\ \textbf{Energy stored in an inductor} = \frac{1}{2}LI^2$$

Worked examples on chapter 4

1 An aeroplane of wing span 40 metres is flying horizontally at 200 ms^{-1}. If the vertical component of the earth's magnetic field is 4×10^{-5} T, calculate the e.m.f. induced between the wing tips.
Area swept out per second by the winds = $200 \times 40\ m^2$
Since $\phi = BA$, the flux cut in one second = $4 \times 10^{-5} \times 200 \times 40$ Webers.
But since the e.m.f. is the rate of cutting of flux
$\therefore$ e.m.f. = $4 \times 10^{-5} \times 200 \times 40$ = **0·32 volt**

2 An 800 turn solenoid has a cross-sectional area of $10^{-3}\ m^2$ and produces a flux density of 4×10^{-3} T when a current of 2 A is passed through it. Calculate its self inductance assuming that the flux density is constant throughout its length.
Flux linking the coil is $\phi = NBA = 800 \times 4 \times 10^{-3} \times 10^{-3}$
$= 3{\cdot}2 \times 10^{-3}$ Wb

Since $e = -d\phi/dt = -L\frac{dI}{dt}$, hence $L = \frac{d\phi}{dI}$

When the current is turned off, the flux will reduce to zero and a change of 2 A in current hence produces a flux change of $3{\cdot}2 \times 10^{-3}$ Wb

$$\therefore\ L = d\phi/dt = \frac{3{\cdot}2 \times 10^{-3}}{2} = \mathbf{1{\cdot}6 \times 10^{-3}\ H}$$

3 A circular coil of 1500 turns and radius 5 cm is rotating at 2000 revolutions per minute about an axis at right angles to a magnetic field of flux density 3×10^{-2} T. What is the peak value of the induced e.m.f.?
The e.m.f. induced is given by $e = NBAw \sin\theta$ and since the maximum value of the sine function is unity, the peak value of the e.m.f. is given by:

$$e\ \text{peak} = NBAw$$

2000 revs min^{-1} = 33·3 revs sec^{-1} so $w = 2\pi F = 2\pi \times 33{\cdot}3$
$\therefore\ w = 66{\cdot}6\ \text{rad sec}^{-1}$
$\therefore\ e\ \text{peak} = NBAw = 1500 \times 3 \times 10^{-2} \times \pi \times (5 \times 10^{-2})^2 \times 47\pi$
$\therefore\ e\ \text{peak} =$ **75 volts**

Practice questions

1 Describe the construction of a simple form of alternating current transformer.
If the primary coil has 1000 turns and is connected to a 240 volt A.C. supply, calculate the current which will flow through a 5 ohm resistor connected to the secondary coil which has 100 turns. (Assume perfect flux linkage and that there are no other losses.)

2 An earth inductor has 200 turns each of radius 20 cm and a resistance of 30 ohms. When the coil is set up in a vertical plane and turned sharply about a vertical axis through 180°, a ballistic galvanometer connected in series gives a throw of 80 divisions. Calculate the measured value of the horizontal component of the earth's magnetic field if the galvanometer has a resistance of 30 ohms and a sensitivity of 10 divisions per microcoulomb.

3 In a post office box, the coils are wound non inductively. Explain what this means and state the reason for using this type of winding.

5 ELECTROSTATICS

Introduction

Simple experiments with charged rods which are not discussed here can be performed to show that there are two charge types namely positive and negative and that **like charges repel each other** whilst **unlike charges attract**.

Traditionally a positively charged rod was produced by rubbing glass with silk and negatively charged by rubbing ebonite with fur though it is more convenient and reliable to use an ordinary duster and a cellulose acetate rod for positive charge and a polythene rod for negative charge. Neither cellulose acetate nor polythene were, of course, available when the early electrostatics experiments were being performed.

The gold leaf electroscope

This was one of the earliest methods of testing charge signs is still used today.

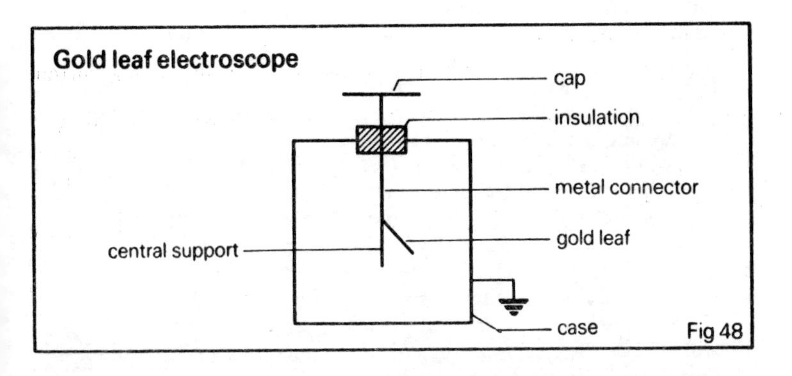

Fig 48

A gold leaf is fitted to a metal rod which is attached to a circular disc or cap. The leaf and rod are screened from outside influence by a transparent case so that the only way to influence the leaf is via the cap.

Consider a negatively charged rod being touched onto the cap. Some of the charge will pass to the leaf and central support, via the metal connector, and the leaf will be repelled away from the central support since like charges repel. **The leaf is said to diverge.**

If an unknown charge is now brought close to the cap and the leaf diverges further this implies that the unknown charge is also negative and repelling more negative charge to the leaf hence making it rise. This method of charging the electroscope is called **charging by contact** and is not as reliable as the method of charging by induction which is discussed below.

Charging an electroscope by induction

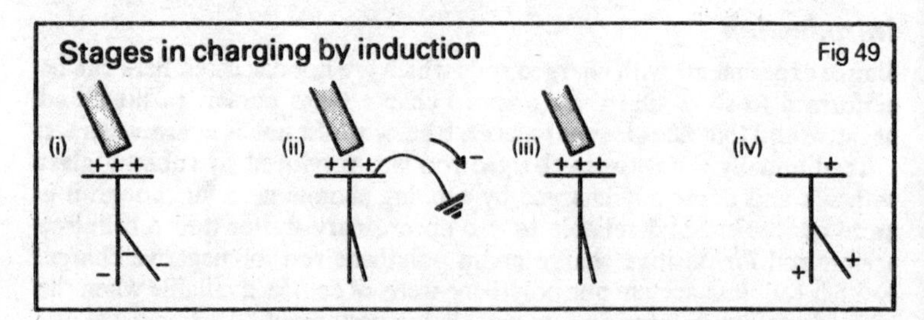

1 A negatively charged rod is brought close to, but not touching, the cap of an uncharged electroscope.
N.B. The term 'uncharged electroscope' does not mean that the electroscope possesses no charges but rather that the positive and negative charges are equal in number and therefore neutralise each other leaving no resultant charge.
As shown in Fig. 49(i) the negative rod in close proximity to the electroscope cap produces a charge separation in the electroscope. Since unlike charges attract and like charges repel, positive charge is attracted to the cap by the negative rod whilst negative charge is repelled to the leaf and central support causing the leaf to diverge.
N.B. Charges **do not** jump the gap between rod and electroscope cap at any stage. There is no actual charge transfer between rod and cap since the two are never brought into contact.

2 The cap is earthed (Fig. 49(ii)). This causes the negative charge to be repelled to earth whilst the positive charge is held on the cap by the negative rod which is still in position. The leaf now falls since there is now no electrostatic force causing it to be repelled.

3 The earth is removed with the negative rod still in position. This prevents the negative charge from returning from earth and leaves the electroscope with a permanent positive charge. This positive charge is still held on the cap by the negatively charged rod and the gold leaf is therefore still in the fallen position.

4 The rod is removed. This allows the positive charge to redistribute itself (like charges repel). The leaf and central support therefore both acquire positive charges and the leaf rises.

N.B. In this method of **charging by induction** the **negatively charged rod** has **induced a positive charge onto the electroscope** without touching it at any time. The electroscope acquires a charge sign which is opposite to that of the rod charging it whilst in charging by contact the charge sign is the same.

Result table

Unknown charge signs can be established by first giving the electroscope a known charge and then bringing the unknown charge close to, but not touching, the cap. The following results are observed:

Charge on electroscope	Charge on rod brought close	Effect on leaf divergence
+	+	increases
−	−	increases
+	−	decreases
−	+	decreases
+	uncharged	decreases

Consider a positively charged rod being brought towards a positively charged electroscope. As the rod approaches, positive charge on the cap of the electroscope will be repelled towards the leaf which will therefore rise further, as it is repelled away from the central support by the extra charge which it now carries.

The **only certain** way to **test a charge sign** is to **obtain increased divergence** since a decrease could mean that the object under test is uncharged.

As well as testing charge signs the electroscope can also measure the distribution of charge over a surface by noting the divergence when charge is sampled from different parts of the surface using a proof plane.

The force between two charges

It can be shown that the force (F) between two charges of magnitude Q_1Q_2 a distance r apart is given by:

$$F \propto \frac{Q_1Q_2}{r^2} \quad \text{the inverse square law}$$

The force is, therefore, directly proportional to the magnitude of the charges and inversely proportional to the square of their distance apart.

$$\therefore F = \frac{KQ_1Q_2}{r^2} \quad \text{where K is a constant}$$

[This expression compares with Newton's Gravitational law: $F = Gm_1m_1/r^2$ for the force between two masses m_1m_1 but the constant G is many orders of magnitude smaller than the electrostatic constant.]

The constant K is usually written as $1/4\pi\varepsilon_0$ where ε_0 is the permittivity of free space (assuming the charges are in a vacuum) and we therefore write:

$$F = \frac{1}{4\pi\varepsilon_0}\frac{Q_1Q_2}{r^2}$$

Permittivity

Expressing the constant as $1/4\pi\varepsilon_0$ may appear to be an unwanted complication but there are, in fact, advantages in putting the constant ε_0 into the denominator of the expression and introducing the extra factor 4π.

Units:

Since $\varepsilon_0 = Q_1Q_2/4\pi Fr^2$ the units are therefore coulomb^2 newton^{-1}

metre^{-2} ($C^2 N^{-1} m^{-2}$) though this is more usually expressed as farad metre^{-1} ($F m^{-1}$). (See later.)

Since $\varepsilon_0 = 8{\cdot}854 \times 10^{-12} C^2 N^{-1} m^{-2}$ the numerical value of the constant $1/4\pi\varepsilon_0$ is about 9×10^9, the significance of this being that it is a large number indicating that the electrostatic force can be quite strong even between relatively weak charges. Comparison with the gravitational constant G shows many orders of magnitude difference indicating why electrostatic forces are relatively easy to detect whilst gravitational forces are not.

If the charges are not in a vacuum then the expression must be modified to:

$$F = \frac{1}{4\pi\varepsilon} \frac{Q_1 Q_2}{r^2}$$

where ε is the permittivity of the medium between the charges

Since ε for air is approximately equal to ε_0 (the free space value) it is usual to use ε_0 in expressions unless a special medium is deliberately inserted between the charges.

Relative permittivity

Unit none Symbol ε_r

The relative permittivity or dielectric constant of a material is the ratio of the permittivity of the substance to the permittivity of free space.

$$\varepsilon_r = \frac{\varepsilon}{\varepsilon_0}$$

ε_r is a dimensionless quantity since it is the ratio of two permittivities which clearly have the same units.

Electric fields

An electric field exists in a region where an electric force will be experienced by a charge and can be plotted using electric field lines in much the same way as magnetic field lines trace the field between two magnets.

The direction of the field is the direction of the force experienced by a positive charge placed in the field and is indicated by an arrow on the field line. The lines of force which are also called **electric flux** therefore start on positive charges and end on negative ones.

Some common electric field configurations are shown in Fig. 50.

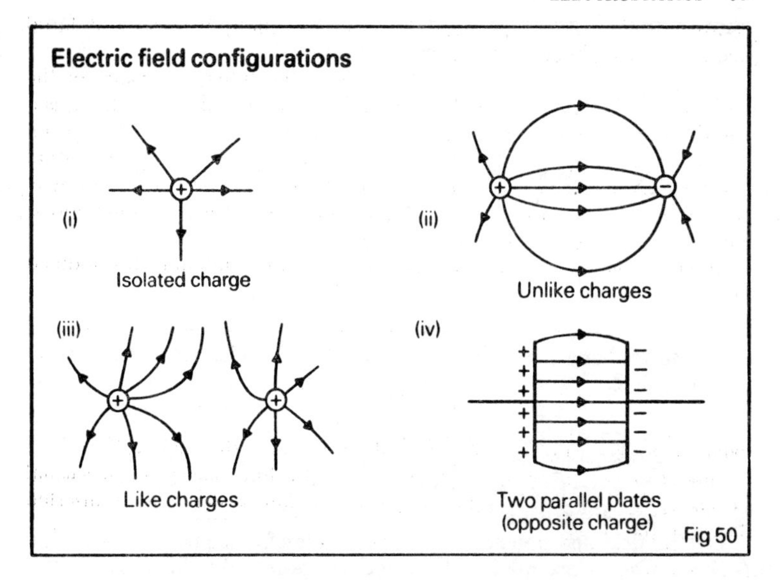

Fig 50

Intensity of an electric field

Unit Newton per coulomb (NC^{-1}) Symbol E

The intensity at a point in an electric field is defined as the force per unit charge acting on a positive electric charge placed at that point.

i.e. $E = \frac{F}{Q}$ where F is force and Q is the charge

This is better expressed as:

$$F = QE$$

Unit: Since E = F/Q the unit of electric field intensity is Newton per coulomb ($N\,C^{-1}$). Sometimes expressed in volts per metre (Vm^{-1}). (See later.)

Electric potential

The concept of electrical potential represents an alternative way of describing charge motion. Consider in Fig. 51 two plates with a potential difference (produced by an E.H.T. supply) of 6000 volts between them. Rather than always discussing a potential difference it is easier to discuss the potential of one plate by referring it to a reference (zero) potential which is often earth in practical cases. For convenience, therefore, we assign a potential of 6000 volts to one volt and zero to the other.

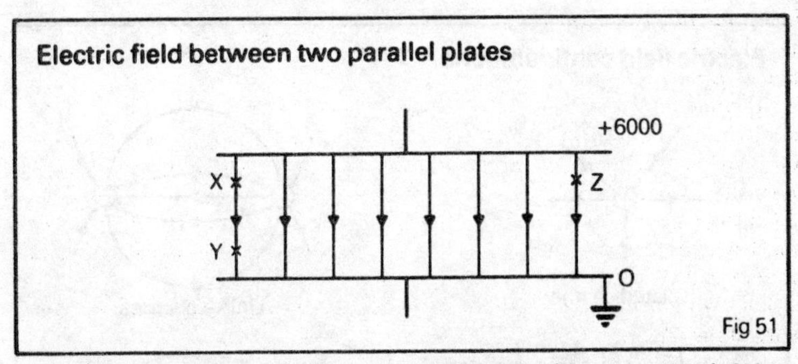

Fig 51

A positive charge placed in the electric field at X will move towards Y but not towards Z. This is because there is a difference in potential (a potential gradient) between X and Y but no gradient between X and Z which are at the same potential since they are equidistant from the +6000 volt plate.

The electric field (as shown by the field line) will move a positive charge from X to Y where there is a potential gradient but there is no motion from X to Z where there is no field and no potential gradient either.

Electric field and potential gradient are therefore equivalent ideas but **field is a negative potential gradient** since the potential is decreasing in the positive direction of the field.

Electric field is equivalent to a negative potential gradient

Equipotential surface

Since X and Z are at the same potential, a line drawn between them will be an equipotential line and there will be an equipotential surface if the argument is extended to three dimensions. Fig. 51 could therefore be extended to show the equipotential lines as well as the electric field lines.

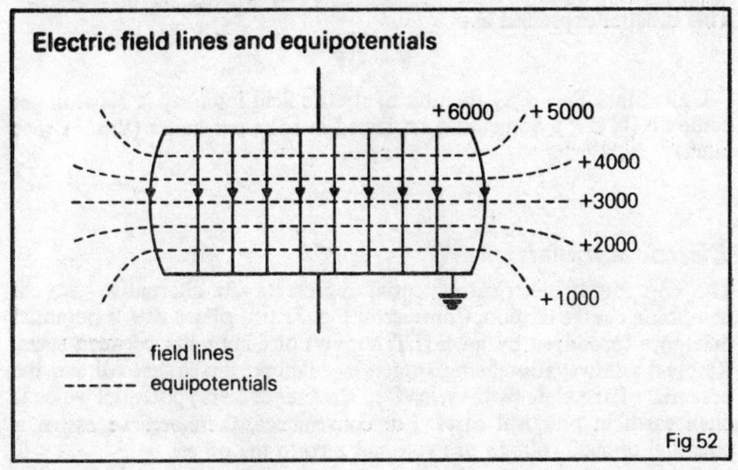

Fig 52

Definition of electric potential

The electric potential at a point is the work done in moving a unit positive charge from the point to a point of zero potential.

(In practical cases zero is usually earth but for theoretical calculations the zero is often taken as infinity because the electric field due to a charged body and earthed conducting surface is complicated resulting in difficult potential calculations.)

Consider moving a charge Q between two points A and B a distance δr apart where δr is so small that the force F produced by the electric field present can be considered constant over this small region.

Since work done is given by the product of force and distance moved, the work done δW is given by:

$$\delta W = F\delta r = QE\delta r \qquad \text{(since } F = QE\text{)}$$

If the difference in potential between the two points is δV we can write:

$$\delta V = \frac{\text{work done by the field}}{Q} = \frac{-\delta W}{Q}$$

(the minus sign since the potential falls in the positive field direction)

$$\therefore \ \delta V = \frac{-QE\delta r}{Q} = -E\delta r$$

$$\therefore \ E = \frac{-\delta V}{\delta r} = \frac{-dV}{dr} \qquad \text{in the limit as } \delta r \to 0$$

$$\therefore \ \mathbf{E} = \frac{-\mathbf{dV}}{\mathbf{dr}} \qquad \text{field is a negative potential gradient}$$

Field of a charged particle

Consider a point charge Q and a test charge Q_1 being placed in the electric of Q a distance r away. The force (F) between the two charges is given by:

$$F = \frac{1}{4\pi\varepsilon_0}\frac{QQ_1}{r^2} \qquad \text{the inverse square law}$$

Also by the definition of field intensity (E) we have:

$$E = \frac{F}{Q_1} = \frac{1}{4\pi\varepsilon_0}\frac{QQ_1}{r^2} = \frac{Q}{4\pi\varepsilon_0 r^2}$$

$$\therefore \ E = \frac{Q}{4\pi\varepsilon_0 r^2}$$

the electric field from a point charge falls off according to the inverse square law ($1/r^2$)

$$\text{Since } E = \frac{-dV}{dr} \qquad \text{(field is a negative potential gradient)}$$

$$\therefore V = -\int_{\infty}^{r} E dr = \int_{\infty}^{r} \frac{Q}{4\pi\varepsilon_0 r^2} dr$$

$$= \frac{Q}{4\pi\varepsilon_0} \int_{\infty}^{r} \frac{1}{r^2} dr$$

$$= \frac{-Q}{4\pi\varepsilon_0} \left[-\frac{1}{r} \right]_{\infty}^{r}$$

$$= \frac{Q}{4\pi\varepsilon_0 r}$$

$$\therefore \mathbf{V} = \frac{\mathbf{Q}}{4\pi\varepsilon_0 \mathbf{r}}$$

The potential decreases according to a $1/r$ law.

N.B. The electric field decreases obeying a $1/r^2$ law and is a vector quantity whilst potential decreases obeying a $1/r$ law and is a scalar quantity.

Field and conductors – some general results

1 The electric field is zero inside a conductor. This must be true since if a field existed it would redistribute the charges present in the conductor until such time as the field no longer existed.
The potential of a conductor is therefore everywhere the same.
If the field is zero then the potential gradient is zero implying that the potential is constant (though not necessarily zero).

2 The field at the surface of a conductor is everywhere normal to the surface. Since field is a vector quantity then, if the field lines were not normal to the surface of the conductor, there would be a component of the field parallel to the surface and this would set the surface charges in motion.
The surface of the conductor is therefore an equipotential surface since electric field lines are normal to equipotentials and we have just argued that the field lines must be normal to the conductor surface. The field lines drawn in all the diagrams so far should, therefore, leave the conductor surface normally.

3 **The charge on a conductor is all on its surface.** If charge existed in the centre then there would be field lines associated with these charges and we have already argued that this is not possible.

4 **The field is zero inside a hollow conductor.** We have already argued that there is no field on the inside of a solid conductor. Hollowing out the middle of the conductor will not change the surface charge distribution and so the field in the centre will still be zero.

5 The field outside a charged conducting sphere is the same as if the charge on it were all concentrated at the centre. (The field inside is zero.) The advantage of this concept is clearly that it allows us to deal with a single point charge at the centre rather than a distribution of charges on the surface.

6 The field on the surface of a conductor is proportional to the charge density on it.
The charge density (ρ) is the charge per unit area so if a sphere carries a charge Q and has radius r then:

$$\rho = \frac{Q}{4\pi r^2} \qquad \text{(surface area} = 4\pi r^2)$$

The field at the surface of the sphere is the same as if all of the charges were concentrated at the centre and is given by:

$$E = \frac{Q}{4\pi\varepsilon_0 r^2} = \frac{4\pi r^2 \rho}{4\pi\varepsilon_0 r^2} = \frac{\rho}{\varepsilon_0} \qquad \text{(since } Q = 4\pi r^2 \rho)$$

E is therefore proportional to ρ since ε_0 is a constant showing that field intensity is directly proportional to charge density.

7 The field (and hence charge density) is greatest at points. The potential V of the sphere discussed above is constant, equal to the potential at the surface and given by:

$$V = \frac{Q}{4\pi\varepsilon_0 r} \qquad \text{– a } \frac{1}{r} \text{ dependence giving}$$

$$Q = 4\pi\varepsilon_0 rV$$

Hence $$\rho = \frac{Q}{4\pi r^2} = \frac{4\pi\varepsilon_0 rV}{4\pi r^2} = \frac{\varepsilon_0 V}{r}$$

$\therefore$ ρ is proportional to $1/r$ since $\varepsilon_0 V$ are constants.

Hence as the radius r is reduced, the object becomes more pointed and ρ increases. The field intensity E therefore increases since it is proportional to the charge density $\rho (E \propto \rho)$.
This fact is used in the electrophorus to draw a spark from a point.

The electrophorus

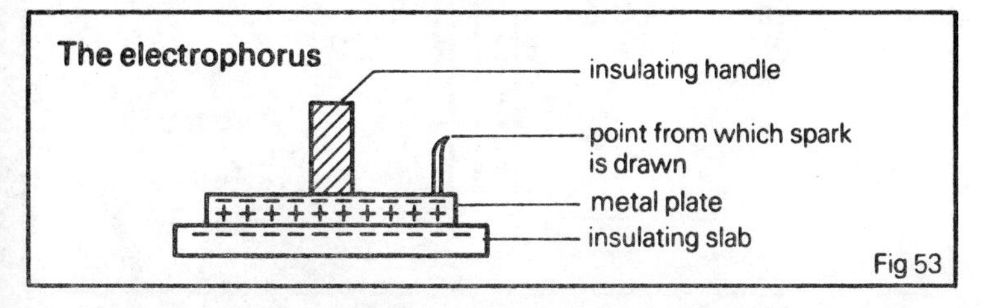

Fig 53

The insulating polythene slab is charged by rubbing it with a duster and the metal plate is placed on it. The negatively charged slab causes the charges in the metal to separate, the positive charge being attracted towards the negative slab whilst the negative charge is repelled towards the top surface of the plate.

Only a small amount of negative charge will flow from the polythene slab to the plate since the two surfaces will, in fact, only touch in a few places since they will not be perfectly flat. Little charge will flow at these contact places since polythene is an insulator.

The metal plate is now earthed allowing the negative charge to be repelled to earth. The earth is now removed so that the negative charge cannot return to the plate which has now been charged positively by induction. The plate can now be lifted off the polythene slab and a visible and audible spark can now be drawn from the point attached to the plate.

An almost unlimited supply of sparks can be produced by repeating this process since the charge on the polythene slab is not used up. Theoretically the process could be continued indefinitely though in practice the charge on the slab eventually leaks away.

The energy for the spark is produced by the work done in lifting the positive plate against the force of attraction from the negative slab. The device is therefore an energy converter and not an energy source as might have been thought at first sight.

The Van De Graaff generator

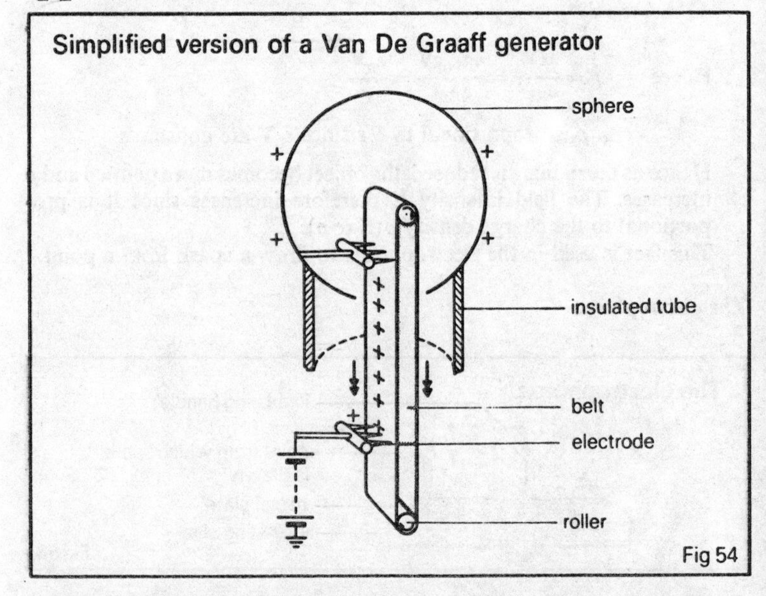

Fig 54

This machine continuously separates charges by induction and is able to build up potential differences of order one million volts on the sphere.

A hollow metal sphere is supported on an insulated tube inside which a silk belt runs over two rollers, the lower roller being driven electrically. Pointed electrodes are arranged close to the rollers as shown in the diagram.

The lower electrode is made about 10,000 volts positive with respect to earth and the high electric field at the points ionises the air. Positive charges are sprayed onto the belt which carries them up towards the sphere.

Negative charge is induced onto the points of the upper electrode and positive charge onto the sphere. The electric field at the points of the upper electrode ionises the air and negative charges are repelled onto the belt which is, therefore, discharged just before it passes over the roller.

The electrical energy comes from the work which the motor has to do in driving the positively charged belt towards the positively charged sphere. As the process continues the sphere gradually charges up but eventually a stage is reached where the charge leaks away as fast as it can be replaced and the potential difference will rise no higher.

Large machines of this type are used in particle accelerators and high-voltage X ray tubes to quote two examples.

Worked examples on chapter 5

1 Calculate the force which would act between two charges of 1 coulomb placed 1 metre apart in a vacuum

$$(\varepsilon_0 = 8{\cdot}84 \times 10^{-12}\,C^2\,N^{-1}\,m^{-2}).$$

$$\text{Since } F = \frac{1}{4\pi\varepsilon_0}$$

$$\frac{Q_1Q_2}{r^2} = \frac{1}{4\pi \times 8{\cdot}84 \times 10^{-12} \times 1}$$

$$= 9 \times 10^9\,N$$

(This figure is exceptionally high because of the small value for ε_0 which appears in the denominator of the expression.) Strong electrostatic forces exist between charged particles.

2 Two positive point charges of 10 micro coulombs and 5 micro coulombs are 20 cm apart in a vacuum. Find the work which would have to be done to halve their separation ($1/4\pi\varepsilon_0 = 9 \times 10^9$). Suppose that the 10 micro coulomb charge is in a fixed position and that the 5 micro coulomb charge is moved a distance of 10 cm towards it.

Since $V = Q/4\pi\varepsilon_0 r$ the potential difference between the two points is given as:

$$V = \frac{Q}{4\pi\varepsilon_0}\left(\frac{1}{0{\cdot}1} - \frac{1}{0{\cdot}2}\right) = 10 \times 10^{-6} \times 9 \times 10^9\,(10 - 5)$$

$$\therefore\ V = 450 \times 10^3 = 450{,}000 \text{ volts}$$

Since 1 volt is 1 joule per coulomb, the work done is:

$$W = QV = 5 \times 10^{-6} \times 450{,}000$$

(using $Q = 5 \times 10^{-6}$ coulombs for the moving charge)

$$\therefore W = 2{\cdot}25\ J$$

3 An isolated conducting sphere of radius 10 cm carries a charge of 1×10^{-7} coulombs. Calculate the electric field intensity and the potential in the following positions:
a On the surface of the conductor, b at a distance 20 cm away from the centre, c at a distance 5 cm away from the centre.
$(1/4\pi\varepsilon_0 = 9 \times 10^9)$
a On the surface:

$$\text{Since } E = \frac{Q}{4\pi\varepsilon_0 r^2} \quad \text{and} \quad V = \frac{Q}{4\pi\varepsilon_0 r}$$

$$\therefore E = \frac{1 \times 10^{-7} \times 9 \times 10^9}{(0{\cdot}1)^2} \quad \text{and} \quad V = \frac{1 \times 10^{-7} \times 9 \times 10^9}{0{\cdot}1}$$

$$\therefore E = 9 \times 10^4\ NC^{-1} \quad \text{and} \quad V = 9 \times 10^3 \text{ volts}$$

b 20 cm away:

$$E = \frac{1 \times 10^{-7} \times 9 \times 10^9}{(0{\cdot}2)^2} \quad \text{and} \quad V = \frac{1 \times 10^{-7} \times 9 \times 10^9}{0{\cdot}2}$$

$$\therefore E = 2{\cdot}25 \times 10^4\ NC^{-1} \quad \text{and} \quad 4{\cdot}5 \times 10^3 \text{ volts}$$

Since $E \propto 1/r^2$ but $V \propto 1/r$, doubling the distance reduces the field intensity by a factor of 4 but only reduces the potential by a factor of 2.

c 5 cm away:
This is inside the sphere itself (radius 10 cm). The field is therefore zero and the potential constant and equal to the value on the surface.

Practice questions

$(\varepsilon_0 = 8{\cdot}85 \times 10^{-12}\ Fm^{-1})$

1 The electric field intensity on the surface of a sphere is $5000\ Vm^{-1}$. Calculate the charge density at the surface. If the sphere has radius 0·20 m calculate its charge and potential.

2 An electric field is set up by arranging a potential difference of 2000 volts between two plates which are 1 cm apart. A charged oil drop of mass 10^{-14} kg is stationary in between the plates with the electric force counterbalancing the weight of the drop. Calculate the charge on the drop. $(g = 10\ ms^{-2})$.

3 Two equal charges repel each other with a force of 0·2 N when they are placed 1 metre apart in a vacuum.
a Calculate the magnitude of the charges.
b Repeat the calculation assuming that the charges were separated by an insulating liquid with a relative permittivity of 10.

4 Describe the basic principles of the action of a Van de Graaff generator. The high voltage terminal of the generator is a sphere of radius 0·4 m, what is the maximum potential to which it can be raised if electrical breakdown of the air occurs when the electric field intensity exceeds 3×10^6 Vm^{-1}?

6 CAPACITANCE

A capacitor is a device for storing charge and in its simplest form consists of two parallel plates though the design of a practical capacitor would be more complicated than this and is discussed later.

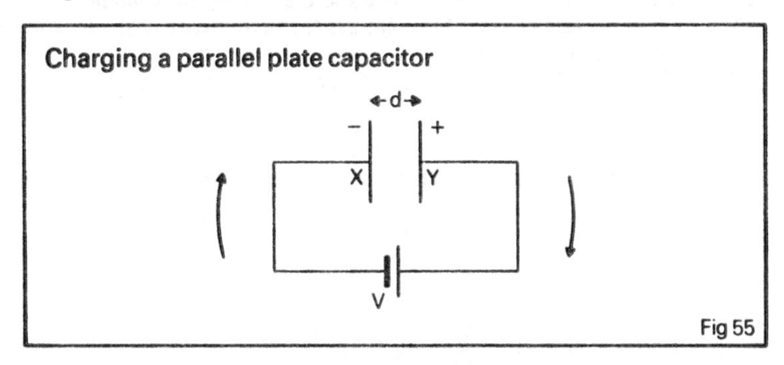

Fig 55

Consider a capacitor being connected across a battery of e.m.f. V volts as shown in Fig. 55. Electrons will flow onto plate X from the negative terminal of the battery (as indicated by the arrows) and towards the positive terminal from plate Y. Plate X will therefore charge negatively whilst Y charges positively.

This flow of charge cannot continue indefinitely since as soon as X begins to charge negatively it will attempt to repel further negative charge away from the plate. Initially the battery will still be able to force more negative charge onto the plate but as the amount of negative charge increases the force of repulsion will also increase and eventually the charge flow ceases. Charge therefore builds up on the plates at an ever decreasing rate and the current eventually falls to zero when the potential difference across the capacitor is equal to the e.m.f. of the battery.

If the battery is disconnected and the two leads are shorted together, a spark will pass as the leads are shorted showing that energy has been stored by the capacitor.

When the battery voltage is increased, more charge (Q) can be stored on the plates and experiments with a ballistic galvanometer (see later) show that V is directly proportional to Q.

$$\text{Since } Q \propto V$$

$$\therefore Q = VC$$

where C is a constant called the capacitance of the capacitor.

Capacitance

Unit Farad (F) Symbol C

Definition: in the parallel plate capacitor the capacitance is one farad if there is a change of one coulomb on each plate when they have a potential difference of one volt between them.

The farad is too large to use as a practical unit and microfarads ($\mu F = 10^{-6}$ F) and picofarads ($pF = 10^{-12}$ F) are more common.

Capacitance of a parallel plate capacitor

Suppose that the plates have an area A and each carries a charge Q. The charge density ρ is given by $\rho = Q/A$ and the electric field intensity E is given by $E = \rho/\varepsilon$ (see chapter 5) where ε is the permittivity of the medium between the plates.

Since field is equivalent to a potential gradient we can write:

$$E = \frac{V}{d}$$

where V is the potential difference between the plates and d is their separation

$$\therefore E = \frac{\rho}{\varepsilon} = \frac{Q}{\varepsilon A} = \frac{V}{d}$$

$$\therefore \frac{Q}{V} = \frac{\varepsilon A}{d}$$

But since Q/V is the capacitance C we can write:

$C = \frac{\varepsilon A}{d}$ the capacitance of a parallel plate capacitor

N.B. 1 Since $\varepsilon = Cd/A$ another possible way of expressing the unit of permittivity is farad per metre (Fm^{-1}) instead of coulomb^2 newton^{-1} metre^{-2} which was used earlier.

2 An alternative unit for electric field intensity is volts per metre (Vm^{-1}) instead of Newton per coulomb (NC^{-1}) which was used earlier.

Methods of increasing the charge stored by the capacitor

Since Q = VC there are two possibilities for increasing the charge stored by the capacitor:

1 Increase the charging voltage. This cannot continue indefinitely since the insulation between the plates of the capacitor will eventually break down if the working voltage is exceeded. This will not only cause a loss of stored energy but will damage the capacitor.
2 Increase the capacitance in three possible ways:
   a Insert a medium between the plates with a higher value of permittivity ε.
   b Increase A – **the area of overlap** of the plates (i.e. build a larger capacitor).
   c Reduce the plate separation d ($C = \varepsilon A/d$). The possible improvement here is once again limited by internal breakdown at very small separation.

Measurement of capacitance using a ballistic galvanometer

Consider the following circuit diagram for the charging and discharging of a capacitor.

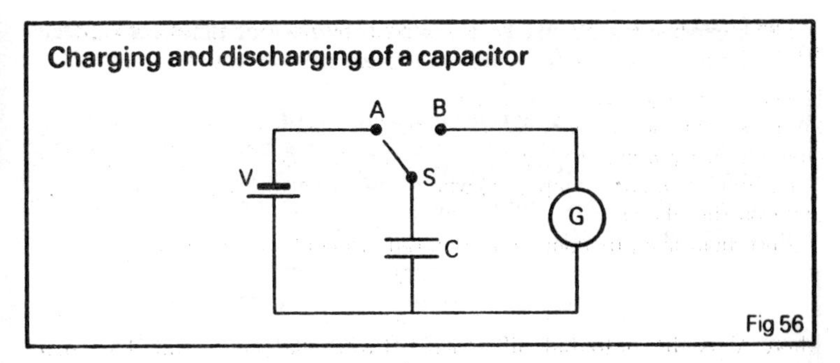

Fig 56

The capacitor C is charged connecting S to A and then discharged through the galvanometer operating ballistically by switching S to B.

If the swing of the galvanometer (θ) is measured for different values of charging voltage (V) it is observed that:

$$V \propto \theta$$

Since

$$Q \propto \theta$$

(theory of the ballistic galvanometer discussed in chapter two where Q is the charge passed)
Hence $Q \propto V$ and Q/V is a constant called the capacitance which is different for different capacitors.

Since

$$Q = VC = K\theta$$

an unknown capacitance can be measured by the following method:

1 Calibrate the system with a known capacitor C_1 and find K by charging the capacitor to a known voltage V_1 and then discharging through the galvanometer to measure the deflection θ_1.

   Since $V_1C_1 = K\theta_1$ the constant is given by $K = \dfrac{V_1C_1}{\theta_1}$

2 Repeat the experiment with an unknown capacitor C and find its capacitance now that K is known. If the deflection on the galvanometer is θ when charged to a voltage V then:

$$VC = K\theta$$

hence:

$$C = \frac{K\theta}{V}$$

Substituting for K then gives:

$$C = \frac{V_1C_1}{\theta_1}\,\frac{\theta}{V}$$

$$= C_1\frac{V_1}{V}\,\frac{\theta}{\theta_1}$$

The experiment should be repeated a number of times for different charging voltages (V) so that an average answer can be obtained.

Measurement of capacitance using a vibrating reed switch

The basic idea is the same as in the previous ballistic galvanometer experiment except that the charge/discharge cycle is extremely rapid (≈ 400 times per second) and is achieved using metal strips (or reeds) which can be magnetised and demagnetised.

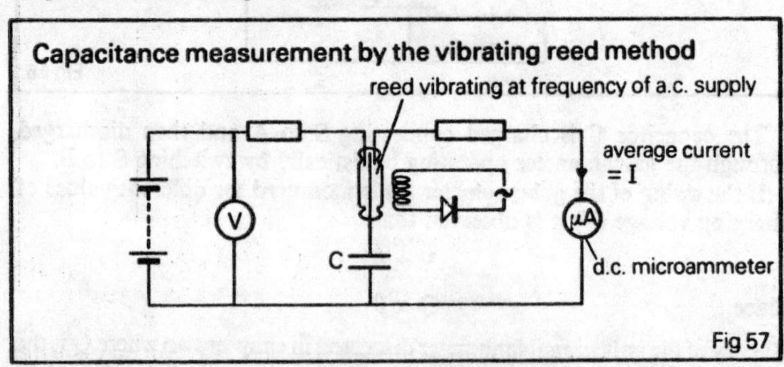

Fig 57

The number of charges and discharges per second is the same as the A.C. frequency driving the reeds and an ordinary meter will show a steady deflection and measure an average current provided the frequency is high enough.

If the frequency of the charge/discharge cycle is f Hertz and each pulse carries Q coulombs of charge then the total charge passing in one second is fQ.

But since $Q = It$ where I is the current we can express current as Q/t (i.e. current is charge passing per second).

Hence $$I = fQ \quad \text{and} \quad Q = I/f$$

Since $$Q = VC \qquad \text{(definition of capacitance)}$$

$$\therefore \frac{I}{f} = VC$$

and so $$C = \frac{I}{fV}$$

The capacitance can therefore be evaluated if the current, charging voltage and frequency are known. Once again a number of readings should be taken by varying f and V and the results should be averaged to obtain the best answer for the capacitance.

Variation of permittivity

Consider inserting an **insulating medium or dielectric** between the plates of a parallel plate capacitor. The charges of the insulator are bound to a particular atom but are capable of a small amount of movement under the action of the electric field between the plates. A small momentary **displacement current** therefore flows whilst the charges in the insulator rearrange themselves and two layers of charge form on the surfaces of the insulator which is said to be **polarised**.

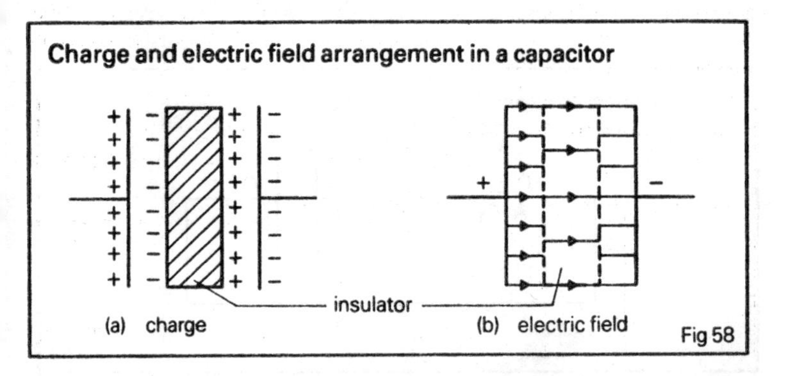

The electric field is set up within the insulator and therefore opposes the original field between the plates and reduces though never eliminates the field in the insulator region. (Fig. 58b.)

If the insulator completely fills the gap between the plates then the field will be reduced and since field is equivalent to a potential gradient this will also be reduced.

Since $$E = \frac{V}{d}$$

where E is the field strength, V is the potential difference between the plates and d is their separation.

It therefore follows that since d is constant then a reduction in E must mean a reduction in V.

Since $Q = VC$ and Q is constant since no charge has flowed on or off the plates, then a reduction in V must mean that the capacitance C has increased.

The capacitor is therefore storing the same charge but at a lower potential difference between its plates. Hence it could store more charge if the potential difference was restored to its original value.

The capacitance is therefore increased by the insertion of an insulating medium between the plates.

This concept can be expressed mathematically by quoting the formula for the capacitance of the arrangement.

$$C = \frac{\varepsilon A}{d}$$

Since the value of permittivity ε for a medium is greater than ε_0 (the approximate air value) it follows that the insertion of the insulating medium will increase the capacitance as discussed above.

Capacitors in series and parallel

a *Series*

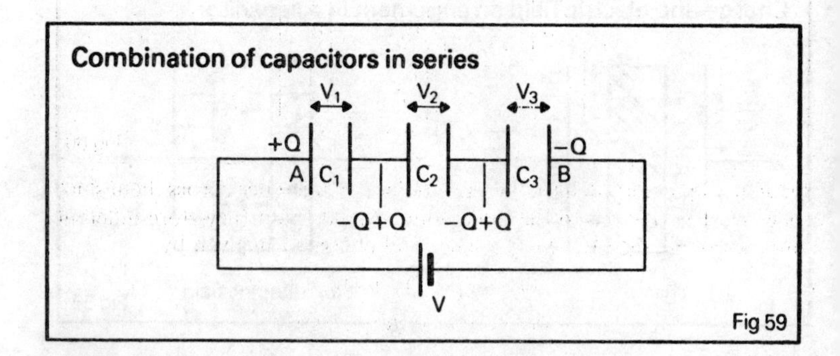

Fig 59

Consider three different capacitors C_1, C_2, C_3 being arranged in series across the terminals of a battery with e.m.f. V volts. The potential differences across the individual capacitors are V_1, V_2, V_3 volts respectively.

If a charge of +Q coulombs is forced onto plate A then −Q will appear on plate B and charges of +Q, −Q will be induced onto all the other plates as shown in Fig. 59. **The same charge is therefore stored by each capacitor** despite the fact that their capacitances are different.

Since the work done in taking a unit charge from B to A is the sum of the work done in taking it across each capacitor in turn we can write

$$V = V_1 + V_2 + V_3$$

Since $Q = VC$ for a capacitor then $V = Q/C$ and hence:

$$\frac{Q}{C} = \frac{Q}{C_1} + \frac{Q}{C_2} + \frac{Q}{C_3}$$

the charge is the same on each capacitor

$$\therefore \frac{1}{C} = \frac{1}{C_1} + \frac{1}{C_2} + \frac{1}{C_3}$$

where C is the capacitance of the combination.

The addition of **capacitors in series** is similar to the addition of **resistors in parallel** and the result of adding another capacitor is to reduce the capacitance of the system.

b *Parallel*

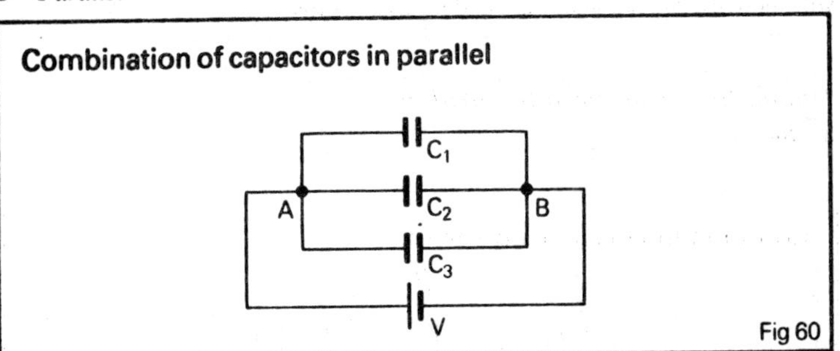

Fig 60

Now the capacitors all have the same potential difference across them since they are all connected to the same points A and B but they store different amounts of charge Q_1, Q_2, Q_3. The total charge Q is given by:

$$Q = Q_1 + Q_2 + Q_3$$
$$\therefore\ VC = VC_1 + VC_2 + VC_3$$
$$\therefore\ C = C_1 + C_2 + C_3$$

where C is the capacitance of the combination.
The addition of **capacitors in parallel** is similar to the addition of **resistors in series** and the result of adding another capacitor is to increase the capacitance of the system.

The energy stored by a capacitor
Consider a capacitor (capacitance C) with a potential difference V between its plates each carrying a charge Q. If we wish to put a small extra amount of charge δQ onto each plate then a small amount of work δW will have to be done. The capacitor will now be storing more energy and the potential difference across its plates will rise by a small amount δV.

If δQ is small the δV will be so small that it is reasonable to consider the potential difference as being constant at the value V.

$$\delta W = VQ \quad \text{(1 volt = 1 joule per coulomb)}$$
$$\therefore\ \delta W = \frac{Q}{C}\delta Q \quad \text{(since } Q = VC\text{)}$$

Hence the total work W done in charging the capacitor from an uncharged state to some specific charge Q is:

$$W = \int_0^Q \frac{Q}{C}\,dQ = \frac{1}{C}\int_0^Q Q\,dQ = \left[\frac{Q^2}{2C}\right]_0^Q = \frac{Q^2}{2C}$$

Since $Q = VC$ we can express the work done and hence the energy stored in three alternative forms:

$$W = \frac{Q^2}{2C} = \frac{QV}{2} = \frac{CV^2}{2}$$

Discharging of a capacitor through a resistor

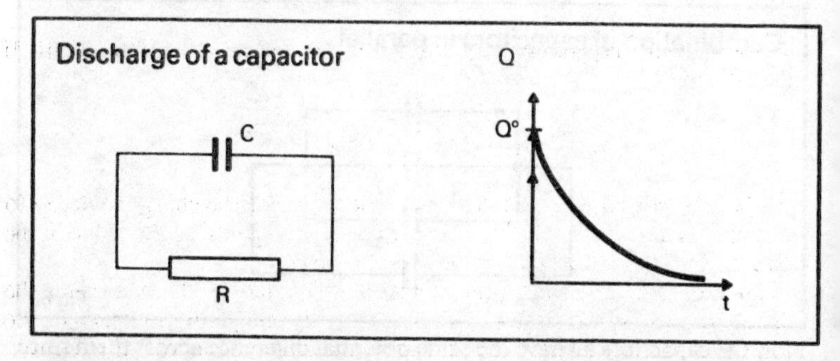

If the capacitor was originally charged to a potential difference of V_0 volts it will have stored a charge $Q = V_0C$ coulombs. Once the discharge has begun this charge will start to flow off the plates and at some time t later, the current I through the resistor will be given by $I = V/R$ where V is the potential difference across the capacitor at time t. The value of the resistance will, therefore, influence the rate of discharge with a high resistance value favouring a slow discharge.

Since $\quad Q = VC \quad$ we have $\quad V = Q/C$

and $\quad I = -dQ/dt$

(current is the rate of flow of charge and the minus sign indicates Q decreasing as time progresses)

The expression $I = V/R$ can therefore be rewritten as:

$$\frac{-dQ}{dt} = \frac{Q}{RC}$$

$$\therefore \int_{Q_0}^{Q} \frac{dQ}{Q} = \frac{-1}{RC} \int_0^t dt$$

where Q_0 is the charge on the plates at time $t = 0$

$$\therefore \ln \frac{Q}{Q_0} = \frac{-t}{RC}$$

$$\therefore Q = Q_0e^{-t/RC}$$

Since the potential difference V is proportional to Q and the current I is proportional to V we can write:

$$V = V_0e^{-t/RC} \quad \text{and} \quad I = I_0e^{-t/RC}$$

where V_0 and I_0 are the initial values of V and I respectively.

The **capacitor** therefore **discharges exponentially** and the time taken for the charge to diminish to half of its original value is always the same irrespective of the charge at the beginning of the time period.

$$e^{-t/RC} = \frac{1}{2} \quad \text{and} \quad t = RC \ln 2$$

The value of the product RC is known as the **time constant** of the circuit. If $t = RC$ then:

$$Q = Q_0 e^{-1} = \frac{Q_0}{e} \text{ when } t = RC$$

The time constant is, therefore, the time taken for the charge to decay to 1/e (approximately 0·37) of its initial value and depends on the value of the resistor and capacitor as is to be expected.

The concept of a time constant is a useful one in characterising the discharge rate of the circuit since the time for complete discharge of the capacitor is theoretically infinite and therefore not practical to measure.

Charging of a capacitor through a resistor

A similar analysis to that carried out for the discharge shows that the charging process is also exponential, the rate of build up of charge depending once again on the resistor and capacitor values chosen.

Practical forms of capacitor

All capacitors consist of two conducting sheets separated by an insulating layer and the following design factors merit consideration.

1 Large areas are needed to produce large capacitance ($C = \varepsilon A/d$). Since physically large components are usually unsuitable, the design often incorporates large area with small volume. Two obvious possibilities are:
   a To use metal foil for the plates and to separate them by an insulator which can be rolled up with the foil.
   b To use a number of plates layered on top of each other and separated by insulating layers.
2 To use thin dielectric layers so that the plate separation (d) can be kept to a minimum.
3 Use an insulator with a high dielectric constant.

The following types of insulator are in common use in capacitors:

a **Paper** usually impregnated with wax to improve the insulation properties. The main advantage is cheapness and physically small size since rolling greatly reduces the volume. Large tolerance values have to be accepted as the capacitance of the device varies with age.

b **Mica** which can be split into very thin uniform sheets typically 0·05 millimetres thick. Layers of metal foil and mica are stacked on top of each other to give a more stable capacitor than the paper type.

c **Ceramic** A wide range of material is available and is compressed into a pipe shape which can then be metallised on the inside and outside to form the electrodes.

Most capacitors show an increase in capacitance with temperature rise but some ceramic capacitors exhibit a negative temperature coefficient and an arrangement of an ordinary capacitor in parallel with a ceramic can produce a capacitance which is temperature independent.

d **Electrolytic** When ammonium borate is electrolysed with aluminium electrodes a thin oxide film forms on the anode in a process known as **anodisation.** This oxide strength is the dielectric in the capacitor and its insulation strength is exceptionally high with the advantage that very thin layers can be used (10^{-7} metres is typical).
High capacitance is therefore combined with small physical size in a capacitor where the two plates themselves are the aluminium anode and the electrolyte to which connections are made via the cathode.
Unfortunately a small current must always pass through the capacitor in the correct direction when it is in use or the anode oxide layer breaks down. Care is therefore needed to connect the capacitor into a circuit in the correct sense. One terminal is usually coloured red and needs to be at a positive potential with respect to the other terminal.
A further disadvantage is that the thickness of the insulating film and therefore the capacitance depends on the average working potential difference between the plates of the capacitor and this may vary from circuit to circuit. This problem is, however, of no consequence in many applications which are discussed in later chapters.

e **Air capacitor** Low insulation strength requires large plate separation and since the plates also have to be mechanically rigid the capacitor tends to be physically large. The main advantage is in the construction of a **variable capacitor** by using two sets of interleaved parallel plates where one set can be rotated to change the area of plate overlap and hence change the capacitance.

Worked examples on chapter 6

1 Two capacitors have capacitance values of 6 μF and 3 μF respectively and are joined in series with a battery of e.m.f. 50 volts. When the charging process is completed the capacitors are disconnected from the battery and their like terminals are joined together. Calculate the final charge on each capacitor.

Since the capacitors are in series the resultant capacitance (C) is given by:

$$C = \frac{\text{product}}{\text{sum}} = \frac{6 \times 3}{6 + 3} = \frac{18}{9} = 2\ \mu\text{F}$$

In series the charge on each capacitor is the same and equal to the charge on the single equivalent capacitor.

$$\therefore\ Q = VC = 50 \times 2 = 100\ \mu\text{C}$$

When the terminals are joined together, the capacitors must come to the same potential difference across their plates and the charge will have to be redistributed although the total charge must remain constant.
If the final potential difference is V volts then the final charge is:

$6\,V + 3\,V$ (microcoulombs)

$\therefore\ 100 + 100 = 6\,V + 3\,V$ so $V = \frac{200}{9}$ volts

$\therefore$ Charge on 6 μF capacitor $= \frac{200}{9} \times 6 = \frac{1200}{9} = \frac{400}{3}\,\mu C$

and charge on 3 μF capacitor $= \frac{200}{9} \times 3 = \frac{600}{9} = \frac{200}{3}\,\mu C$

2 Two insulating spheres of radius 5 cm and 10 cm are charged to potentials of 400 volts and 200 volts respectively and then connected together by a wire. Calculate the energy stored by the system before and after the connection is made. ($1/4\pi\varepsilon_0 = 9 \times 10^9$)
For the sphere $Q = VC$ and $V = Q/4\pi\varepsilon_0 r$
Hence replacing Q by $4\pi\varepsilon_0 rV$ gives $4\ \varepsilon_0 rV = VC$ and so

$C = 4\pi\varepsilon_0 r$ – **capacitance of a sphere radius r**

The capacitances are:

Small sphere $C = \frac{5 \times 10^{-2}}{9 \times 10^9} = \frac{5}{9} \times 10^{-11}$ F

Large sphere $C = \frac{10 \times 10^{-2}}{9 \times 10^9} = \frac{10}{9} \times 10^{-11}$ F

Since energy stored is given by $(1/2)\,CV^2$

$\therefore$ Small sphere stores $\frac{1}{2} \times \frac{5}{9} \times 10^{-11} \times (400)^2 = 4{\cdot}4 \times 10^{-7}$ J

Large sphere stores $\frac{1}{2} \times \frac{10}{9} \times 10^{-11} \times (200)^2 = 2{\cdot}2 \times 10^{-7}$ J

$\therefore$ Total energy stored before connecting

$$= (4{\cdot}4 + 2{\cdot}2) \times 10^{-7}$$
$$= 6{\cdot}6 \times 10^{-7}\ \text{J}$$

When the spheres are joined together, the total charge must remain unchanged and the potential difference across the two will be the same. Since $Q = VC$ and $C = 4\pi\varepsilon_0 r$ we have $Q = 4\pi\varepsilon_0 rV$
$\therefore$ Initial charge is:
$4\pi\varepsilon_0\,((5 \times 10^{-2} \times 400) + (10 \times 10^{-2} \times 200))$
If the final potential difference is V volts then the final charge is given by:
$4\pi\varepsilon_0\,(5 \times 10^{-2}\,V + 10 \times 10^{-2}\,V)$
Equating these two expressions and solving eliminates the term in $4\pi_0$ and gives:

$$V = 267 \text{ volts}$$

$$\therefore \text{ Final energy} = \frac{1}{2}C_{small}V^2 + \frac{1}{2}C_{large}V^2 = \frac{1}{2}(C_{small} + C_{large})V^2$$

$$= \frac{1}{2}\left(\frac{5}{9} \times 10^{-11} + \frac{10}{9} \times 10^{-11}\right) \times (267)^2$$

$\therefore$ Final energy $= 5{\cdot}94 \times 10^{-7}$ J

Approximately 10% of the energy is, therefore, apparently lost when the connection is made. This 'missing' energy is converted into heat in the connecting wire since a current flows whilst the potential differences equalise themselves.

3 A circular coil has 300 turns and diameter 25 cm and is connected to a ballistic galvanometer in a circuit where the total resistance is 1200 ohms. When the coil is turned through 180° a maximum reading of 30 scale divisions is produced on the galvanometer.
The galvanometer is calibrated by charging a known capacitor and then discharging it through the meter. If 20 scale divisions are produced by charging a 0·2 μF capacitor to 3 volts, calculate the vertical component of the earth's magnetic field if the coil was horizontal before the measurement commenced.
Since Q = VC the charge stored by the capacitor is:
$Q = 3 \times 0{\cdot}2 \times 10^{-6} = 6 \times 10^{-7}$ coulombs
Hence the charge produced by the search coil rotation is

$$6 \times 10^{-7} - \frac{30}{20}C$$

$\therefore$ Charge passed $= 9 \times 10^{-7}$ C
For the search coil (earth inductor):

$$\text{Induced charge} = \frac{\text{charge in flux linkage}}{r} = \frac{2BAN}{r} \text{ (chapter 4)}$$

$$\therefore B = \frac{Qr}{2AN} = \frac{9 \times 10^{-7} \times 1200}{2\pi \times (0{\cdot}125)^2 \times 300} = 3{\cdot}7 \times 10^{-5}\,T$$

4 A parallel plate capacitor consists of six metal plates each of area 15 cm² separated by sheets of mica 0·1 mm thick and having a dielectric constant of 6. Calculate the capacitance of the arrangement. ($\varepsilon_0 = 8{\cdot}85 \times 10^{-12}\,Fm^{-1}$)

The capacitance of a single capacitor is $C = \dfrac{\varepsilon A}{d} = \dfrac{6\varepsilon_0 A}{d}$

The six plates will produce five capacitors which are arranged in parallel by joining the odd numbered plates together and the even numbered plates together. The total capacitance is, therefore, additive. ($C = C_1 + C_2 + C_3$ for capacitors in parallel.)

$$\therefore C_{total} = \frac{5 \times 6 \times 8{\cdot}85 \times 10^{-12} \times 15 \times 10^{-4}}{2 \times 10^{-4}}$$

$$\therefore\ C_{total} = 4 \times 10^{-9}\,\text{F}$$

Practice questions

1 Define the terms electrostatic potential and capacitance.

2 A parallel plate air capacitor is charged to a potential difference of 500 volts and is then connected in parallel with an exactly similar capacitor except that this second capacitor has a dielectric inserted between its plates. Calculate the relative permittivity of the dielectric if the potential of the combination is found to be 100 volts.

3 Describe a method of measuring the capacitance of an unknown capacitor by using a standard (known) capacitor of similar capacitance.

4 A 3 μF capacitor is charged to a potential difference of 200 volts. The supply is then replaced by an uncharged 5 μF capacitor. Calculate (a) the resulting potential difference between the plates of the capacitors (b) the energy stored in the system before and after the rearrangement was made.

7 THE CATHODE RAY OSCILLOSCOPE

The cathode ray oscilloscope (C.R.O.) operates by firing a beam of electrons across an evacuated tube and onto a fluorescent screen often coated with zinc sulphide.

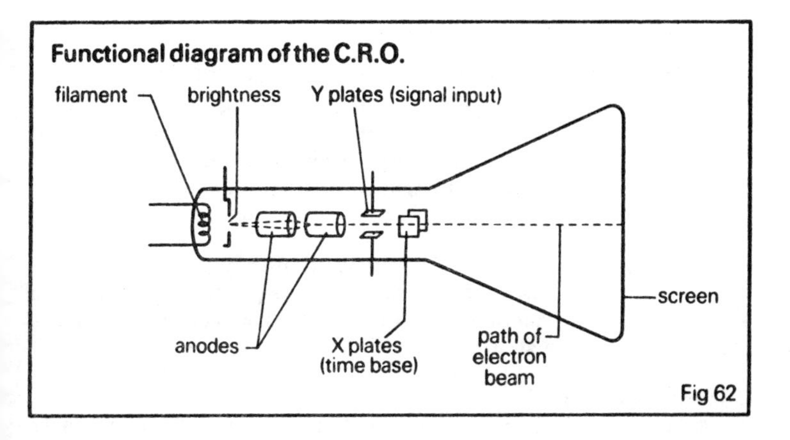

Fig 62

The filament is heated electrically using either an A.C. or D.C. source and electrons are emitted when the filament is hot in a process known as **thermionic emission**. These electrons are attracted by the anode cylinders and many of them pass through and move towards the screen. The anodes are designed to focus the emerging electrons into a narrow beam so that a single spot will be seen on the screen when the electrons strike the zinc sulphide coating and cause it to fluoresce.

Before reaching the screen the electrons pass through two pairs of deflecting plates called the X and Y plates which are capable of deflecting the beam horizontally and vertically when a potential difference is placed across them.

The brightness of the beam is controlled by an electrode next to the filament. As this electrode is made more negative with respect to the filament the number of electrons passing along the tube is reduced and the brightness is, therefore, diminished.

The screen, final anode and the tube wall between the two are usually connected to earth so that the movement of earthed objects such as the operator's hands will not affect the electron path. The filament is therefore held at a negative potential with respect to earth.

The electrons which strike the screen need to return to the anode to complete the circuit otherwise the screen would charge negatively and prevent the arrival of further electrons. This return can be accomplished in two ways:

a The inside of the screen and tube wall to the anode are coated with a conducting layer of graphite.

b When electrons strike the surface they cause the emission of further electrons from it in a process known as **secondary electron emission.** A negative electron space charge therefore builds up by the screen and slowly drifts back towards the anode.

Measurement of an unknown potential difference

If a potential difference is applied across the Y plates then the electron beam will be deflected either up or down depending on the polarity of the input and the spot on the screen will be displaced vertically. If the Y plates have been previously calibrated by using a known potential difference then the unknown potential difference can be measured by noting the deflection on the screen.

Some oscilloscopes have the Y plate input already pre-calibrated at a number of known sensitivities measured in volts per centimetre whilst in similar versions the calibration needs to be carried out before the unknown signal can be measured.

The time base circuit

A constant potential difference could also be applied to the X plates to make the spot move horizontally but it is more usual to apply a sawtooth waveform as shown in Fig. 63.

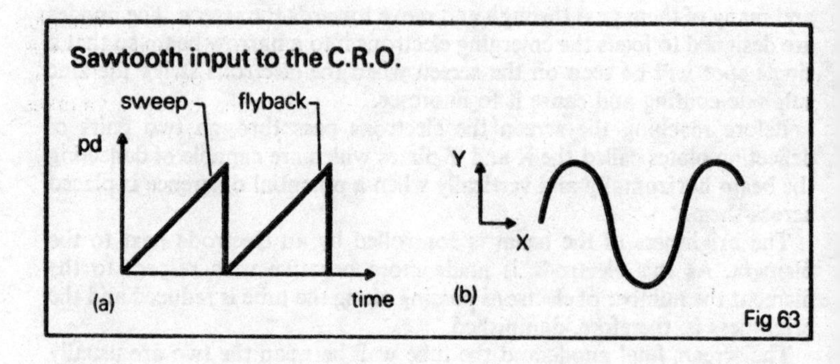

Fig 63

This sawtooth input sweeps the spot across the screen from left to right and then quickly makes it flyback to the starting position ideally in zero time so that the flyback is not observed. The sweep speed can be varied to give a variable time base control which is often calibrated directly so that the sweep speed is known.

If an alternating potential difference were applied to the Y input with the time base off then the spot on the screen would be moved rapidly up and down at the frequency of the input and a vertical line would be seen on the screen. When the time base input is applied to the X plate at the same time as the Y input the spot is pulled across the screen as well as moving up and down and the waveform is traced out as shown in Fig. 63b which shows a sinusoidal signal.

Uses of the cathode ray oscilloscope

1 *As a voltmeter*

When the time base circuit is switched off, a D.C. input to the Y plates will cause a vertical deflection of the spot and an A.C. input will oscillate the spot up and down to produce a vertical line on the screen (unless the frequency is very low when the individual positions of the spot will be visible). Once the Y input of the oscilloscope has been calibrated the spot deflection can be translated into a measurement of the applied potential difference.

The oscilloscope is particularly useful as an **A.C. voltmeter** since the ordinary moving coil meter discussed in chapter 2 will not respond to A.C. **N.B.** The **peak value** of the voltage is measured by the length of the trace on the screen. This is discussed further in the next chapter.

2 *Measurement of frequency*

a **By using a calibrated time base**

This is perhaps best discussed by reference to a specific example. Suppose that the time base is set at '10 millisec/cm' and a sine curve is seen on the screen where the horizontal distance occupied by one complete cycle is 2·0 cm. The time base setting means that the spot takes 10 milliseconds to move one centimetre across the screen.

∴ Time to move 2 cm $= 2 \times 10 = 20$ ms $= 20 \times 10^{-3}$ S

∴ The time to complete one cycle of the input is 20×10^{-3} s

Since frequency = 1/time

$$\therefore \text{ Frequency} = \frac{1}{20 \times 10^{-3}} \text{ 50 Hz}$$

b **By using the method of Lissajous figures**

This method relies on comparing two frequencies by applying one to the *Y* deflector plates and the other to the *X* plates with the time base switched off. A steady, easily recognisable trace is visible on the screen only when the two applied frequencies are in a simple ratio to each other. The traces observed are well documented and some of the simple ratios are shown in Fig. 64.

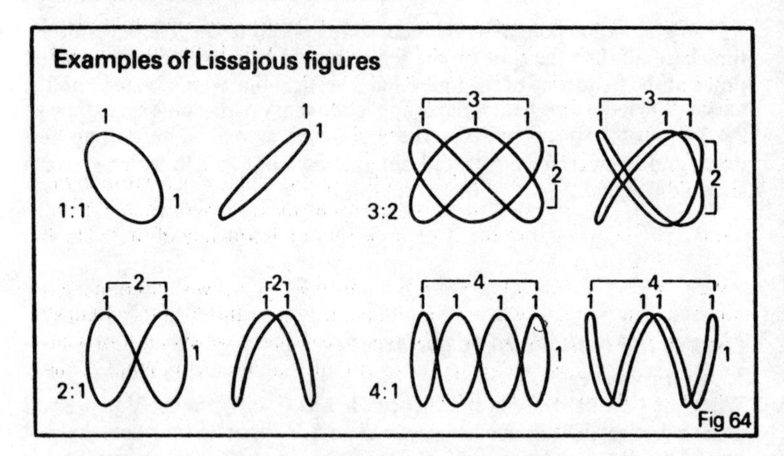

Fig 64

The frequency ratio can be deduced by comparing the number of loops along the top to the number of loops down the side of the observed trace. If one of the frequencies is known accurately then the other one can be found and checked by repeating the comparison for different simple ratios. This method is very accurate since the stationary trace is well defined and can be set up precisely (i.e. it is not observed for a wide frequency range on the standard oscillator).

3 *Measurement of phase*
Two different signals can be easily compared on a double beam oscilloscope to check their phase (i.e. to see if they reach their maxima at the same time).

4 *Observation of the quality of a signal*
A good amplifier would produce a perfect scaled up version of the signal to be amplified although in practice distortion tends to be produced. If the performance of an amplifier becomes unacceptable, the oscilloscope can be used to trace the signal through each of the amplification stages until the distortion is first recognised in the waveform produced. The correction stage of the amplifier can now be repaired to improve the signal quality.

Practice questions

1 Discuss the essential features of a cathode ray oscilloscope using a method of electrostatic deflection.
2 Discuss the use of a cathode ray oscilloscope:
a As a voltmeter, b As a frequency measuring device, c As a method of measuring short time intervals, d As a method of comparing the phase relationship between two different signals.
3 The X and Y plates of a cathode ray oscilloscope are both set to a

sensitivity of 10 volt cm^{-1}. Two different sinusoidal signals both at 50 Hz are applied to the plates, the Y input varying between ± 20 volts whilst the X input varies between ± 30 volt.

Sketch and discuss the patterns observed:

a When the two signals are in phase.

b When they are out of phase by 90°.

4 A voltage alternating at 50 Hz is applied to the Y plates of a cathode ray oscilloscope. Sketch and discuss the waveforms observed when a linear time base connected to the X plates is set at a frequency of: a 25 Hz, b 50 Hz, c 100 Hz.

What trace would be observed if the time base circuit was disconnected and the same signal was fed simultaneously into the X and Y plates?

5 A sinusoidal trace on an oscilloscope screen has a wavelength of 3 cm when the time base circuit is set to run at 5 millisec/cm. What is the frequency of the applied signal?

8 ALTERNATING CURRENTS AND VOLTAGES

Alternating currents and voltages are easier to generate than their direct counterparts and can be transported more conveniently from power station to consumer using transformers to step the voltage up and then down (chapter 4).

The most usual type of alternating waveforms are sinusoidal and can be represented mathematically by:

$$\text{Current} = I = I_0 \sin wt$$
$$\text{Voltage} = V = V_0 \sin wt$$

Where w is the angular frequency and I_0 and V_0 are the maximum values of current and voltage respectively.

Alternating current and voltage is just as efficient as direct for heating and lighting purposes and can be rectified if necessary using diodes (discussed in chapter 10) for applications where an alternating variation is not appropriate.

One problem, however, is finding an average or representative value for the alternating waveform which is constantly changing.

The root mean square (R.M.S.) value

In choosing an average value the mean is clearly not appropriate since it is zero and taking an average over half a cycle only is plausible but does not give a useful result.

The power (P) dissipated in a resistor (R) carrying a current (I) is given by:

$$P = I^2R$$

In an alternating waveform the value of I and hence P is constantly changing but the average power P is given by:

$$\bar{P} = (\text{mean of } I^2) \times R = I_{AV}{}^2R$$

Where $$I_{AV} = \sqrt{\text{mean of } I^2}$$

Since this value I_{AV} is calculated by taking the square root of the mean of the squares it is called the R.M.S. or root mean square value.

This root mean square (R.M.S.) value of the alternating current is the same as that value of steady current which would dissipate energy at the same rate in a fixed resistance. The R.M.S. value is, therefore, the **'D.C. equivalent'** and a light bulb marked 240 volt, 100 watt, for example, will be equally bright whether it is run from a 240 volt A.C. or D.C. source, since 240 volt A.C. would mean 240 volt A.C. (R.M.S.).

Relationship between the R.M.S. and peak value

If the waveform is sinusoidal then the current variation with time is given by

$$I = I_0 \sin wt.$$

The R.M.S. current ($I_{R.M.S.}$) is therefore given by:

$$I_{R.M.S.} = (\text{mean value of } I^2)^{1/2}$$

$$\therefore\ I_{R.M.S.} = I_0\,(\text{mean value of } \sin^2 wt)^{1/2}$$

often written $I_{R.M.S.} = I_0\,(\overline{\sin^2 wt})^{1/2}$

Since $\sin^2 wt = \frac{1}{2}(1 - \cos 2\,wt)$ (standard mathematical formula)

$$\therefore\ I_{R.M.S.} = I_0\left(\frac{1}{2} - \frac{\overline{\cos 2\,wt}}{2}\right)^{1/2}$$

Since the mean value of cos 2 wt is zero, as it is an ordinary cosine function, the expression for $I_{R.M.S.}$ therefore simplifies to:

$$I_{R.M.S.} = I_0\left(\frac{1}{\sqrt{2}}\right)$$

usually written

$$I_{R.M.S.} = \frac{I_0}{\sqrt{2}} = 0{\cdot}707\,I_0$$

The peak value of the 240 volt mains voltage in Great Britain is therefore $240\sqrt{2} \approx 339$ volts though the D.C. equivalent or R.M.S. value of 240 volts is the value normally quoted.

The peak to peak variation which would be measured in a cathode ray oscilloscope is therefore ± 339 volts.

Relationship between R.M.S. and peak value shown experimentally

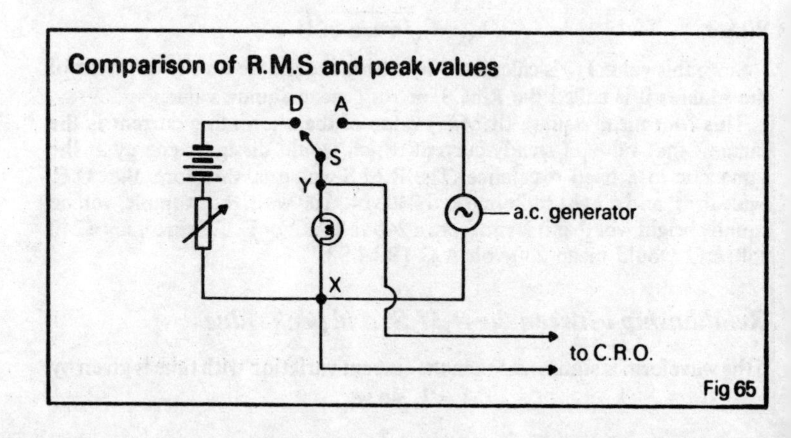

Fig 65

S is a change-over switch so that the light bulb can be illuminated either by D.C. when S is connected to D or by A.C. when S is connected to A. The rheostat is adjusted until the lamp is equally bright whether it is illuminated by A.C. or D.C., the switch being continuously changed between D and A whilst the comparison is made.

This equal brightness measurement can be performed by eye but in preference a photocell should be used. The light bulb and photocell are enclosed in an opaque box so that external room lighting will not interfere with the photocell reading.

The potential difference between Y and X (i.e. across the bulb) can be measured using a cathode ray oscilloscope (chapter 7) when the rheostat adjustment is completed. The deflection of the spot (D.C.) can be compared with the length of the line observed (A.C. with the time base off).

The D.C. reading is measuring the R.M.S. or 'D.C. equivalent' value whilst the A.C. reading is measuring the maximum or peak value. The readings can be repeated for a number of different light intensities but the ratio is always $2\sqrt{2}$.

(Not $\sqrt{2}$ since the C.R.O. measures the peak to peak value of the A.C. which is twice the peak value.)

Alternating current meters

An ordinary D.C. moving coil meter is not suitable for A.C. measurements since at low frequency it would oscillate slowly and at typical mains frequency (50 Hz) and higher it would not be able to respond quickly enough and would merely vibrate slightly about the rest position. Different types of measuring device are therefore required.

a *Rectifier instruments*

These are, perhaps, the most important category and rely on one or more diodes (chapter 10) to rectify the A.C. to D.C. so that the usual moving coil meter movement can be used. A.C. voltmeters and ammeters are usually calibrated to read the R.M.S. value directly.

1 Simple rectifier

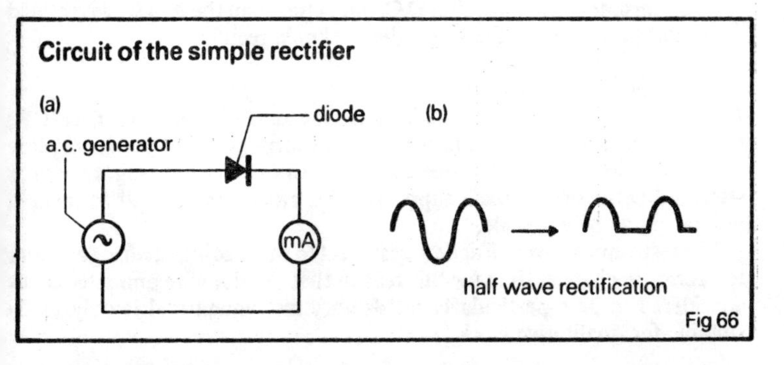

Fig 66

Since the diode will conduct a current in one direction only it conducts in alternate half cycles of the alternating input. The sine wave input is therefore converted to a pulsating D.C. waveform as shown in Fig. 66b. There will be a steady deflection proportional to the half cycle average of mean value of the current provided the frequency is not too low when pulses will be observed.

2 **The diode bridge circuit**

This arrangement (Fig. 67) also uses diodes to rectify the A.C. waveform and consists of a bridge of four diodes with the A.C. generator across one pair of opposite corners and a normal meter movement across the other pair.

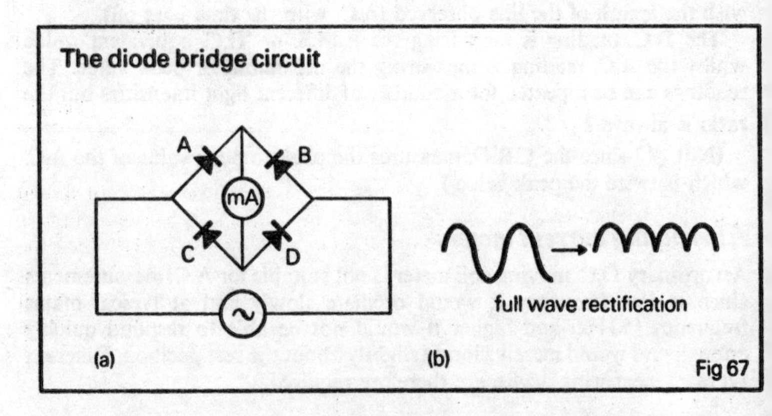

Fig 67

Careful inspection of the diode arrangement shows that diodes **will conduct in pairs** with B and C allowing current to pass during one half cycle whilst A and D will conduct in the other half cycle. The important point is that the **current flows through the meter in the same direction during both half cycles** and the alternating current is, therefore, **full wave rectified** as shown in Fig. 67b.

The current is still pulsating D.C. but better than the half wave rectified signal produced by the single diode simple rectifier.

b *The hot wire ammeter*

In this instrument the pointer deflection depends on the heating effect (I^2R) of an electric current. The current is passed through a fine platinum alloy wire and raises its temperature causing it to expand and sag since it is stretched between two fixed supports. This movement is used to make a pointer move over a scale.

The instrument is very fragile and subject to the readings drifting as room temperature changes. It is for this reason that the hot wire ammeter is not considered to be a particularly satisfactory instrument and is only really suitable for qualitative work.

c *Dynamometer instruments*
These instruments consist of two fixed coils and a moving coil pivoted on bearings.

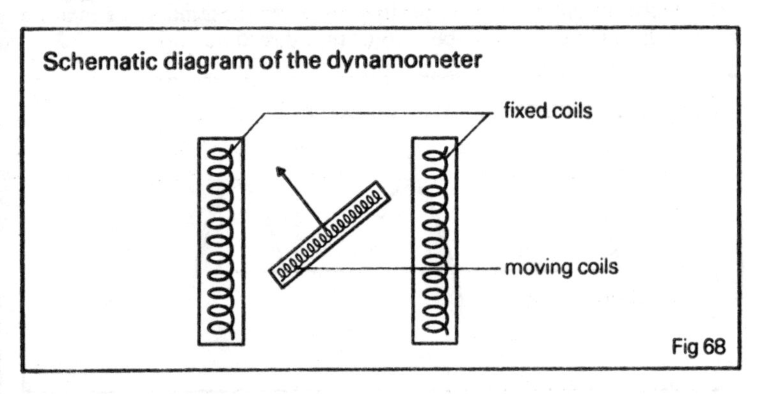

Fig 68

The moving coil also carries a pointer which moves over a scale (not shown on Fig. 68). The coils are joined in series such that a deflecting couple is established by the interaction of the associated magnetic fields. When the current in the coils changes direction the magnetic fields of both the moving coil and the fixed coils will reverse and the deflecting couple is therefore still acting in the same direction. A steady reading is, therefore, obtained on the scale.

d *The cathode ray oscilloscope* (C.R.O.)
Full details are given in chapter 7 but two important differences between this instrument and the other A.C. meters are worth stressing.
1 This voltmeter measures the peak to peak value of the A.C. variation whilst the other instruments are usually calibrated for R.M.S. values.
2 Since a visual display is available the quality of the waveform can also be monitored. This is obviously not possible with the other instruments.

A.C. and resistors, inductors and capacitors

1 *Resistors*
Ohm's law is still valid provided the R.M.S. (D.C. equivalent) values of current and voltage are used. There are no special problems associated with the magnetic field constantly changing provided wire resistors are wound non inductively (chapter 4).

2 *Inductors*
Since the current is changing direction the associated magnetic field is also changing direction and a back e.m.f. is induced which inhibits the current change (as discussed in chapter 4). Suppose that the coil has negligible

resistance so that only the inductive effect need be considered and the problem is simplified.

Consider a plot of current variation against time where points marked X indicate the current flowing in a positive sense and those marked Y indicate a zero where the current is about to change direction.

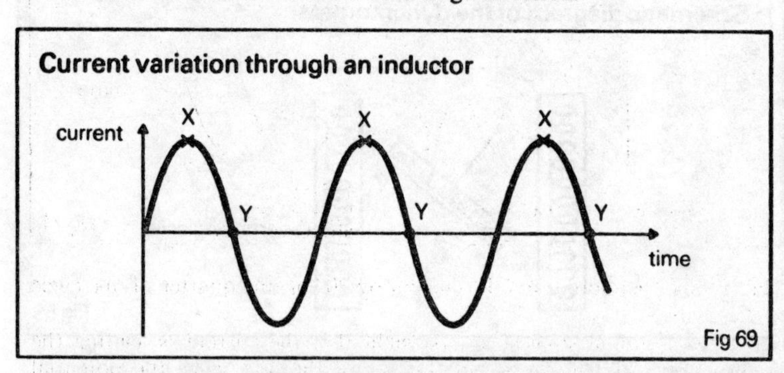

Fig 69

At X the current has reached a maximum and is temporarily not changing (dI/dt = zero at points X). The magnetic field lines are therefore not moving and the back e.m.f. induced will be temporarily zero. The current and the back e.m.f. are therefore out of phase by 90° or one quarter of a cycle.

At Y when the current is zero, the rate of change of current is greatest (steepest gradient) and the back e.m.f. is therefore a maximum and will be in a positive sense so that it can oppose the current which is swinging negatively. Fig. 70 shows the phase relationship between the current and back e.m.f.

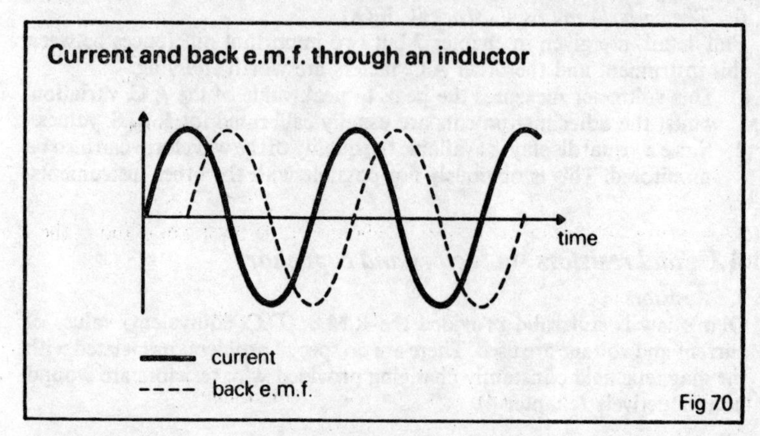

Fig 70

Since the resistance is negligible, the applied potential difference must be equal and opposite to the back e.m.f. if an alternating variation is to be maintained. Fig. 71 shows the phase relationship between the applied potential difference and the current.

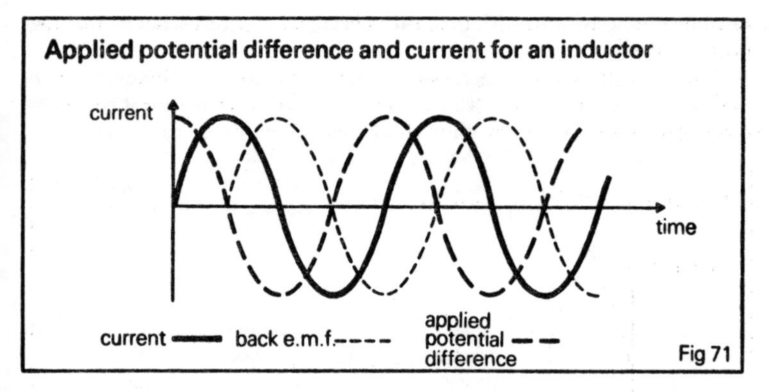

Fig 71

The voltage therefore leads the current by 90° or one quarter of one cycle ($\pi/2$ radians).

N.B. Fig. 71 might suggest at first sight that the current is leading the applied potential difference but this is not the case since the potential difference is reaching its **maximum first in time.**

Inductor V leads I

This phase difference between current and voltage can be demonstrated with an inductor and resistor in series using a double beam oscilloscope as a voltmeter. Input one is placed across the inductor to measure potential difference whilst input two is measuring the potential difference across the resistor but since this is in phase with the current, a measure of the current phase is therefore achieved. Comparison of the two traces on the oscilloscope shows them to be out of phase with each other demonstrating the phase difference between current and voltage for an inductor.

Mathematical derivation of the phase relationship for an inductor

The current variation is given by $I = I_0 \sin wt$.

and $$e = -L\,dI/dt$$

(definition of inductance in chapter 4 where e is the back e.m.f. and L the inductance of the coil).

$$\therefore\ e = -L\frac{d}{dt}(I_0 \sin wt) = LI_0 w \cos wt$$

$$\therefore\ e = -LI_0 w(\sin wt + \pi/2)$$

($\pi/2$ phase difference between sine and cosine functions)

Since the applied p.d. $= -e$ for the inductor

$$\therefore\ \text{Applied p.d.} = LI_0 w\ (\sin wt + \pi/2)$$

The current and potential difference are therefore 90° ($\pi/2$ radians) out of phase with the potential difference leading. This mathematical reasoning is therefore in agreement with the result discussed previously but is obtained rather more quickly.

N.B. 1 Since we can now have $V = 0$ yet $I \neq 0$ for the inductor the use of **Ohm's law** ($V = IR$) in its simplest form is clearly **not now valid.** This basic error must be avoided and a modification is needed using the R.M.S. values as discussed later.

2 The simplest power expression: power = volts × amps (chapter 1) is also not now valid since the power would regularly become zero during the cycle.

Change of frequency

As the frequency of the applied signal increases there is less time for current changes to occur. If the applied potential difference is constant then the back e.m.f. which is in opposition to it must also be constant.

Since $e = -L\,dI/dt = \text{constant}$

Hence the rate of change of current (dI/dt) is also constant.

i.e. $\dfrac{\text{current change}}{\text{time taken}} = \text{rate of change of current} - \text{constant}$

Hence if the time gets less the current must also get less and the current is, therefore, inversely proportional to the frequency.

i.e. $I \propto 1/f$ where f is the frequency of the applied signal

Hence as the frequency is increased the inductive effect gets worse and the reactance (analogous to A.C. resistance) increases.

Reactance

Unit Ohms (Ω) Symbol X

The reactance measures the opposition which an inductor (or capacitor) offers to A.C. and is defined as:

$$\text{Reactance of inductor} = X_L = \frac{\text{Amplitude of voltage across the inductor}}{\text{Amplitude of current through it}}$$

$$= \frac{V_0}{I_0}$$

Since the reactance is a ratio of a voltage to a current it is therefore measured in ohms.

N.B. The term resistance is not used since a resistor dissipates electrical power as heat whilst an inductor and capacitor store energy but do not dissipate it.

Energy stored $= \frac{1}{2}LI^2$ for the inductor (chapter 4)

Energy stored $= \frac{1}{2}CV^2$ for the capacitor (chapter 6)

Since $I = I_0 \sin wt$

and applied potential difference $= V = LI_0 w \sin(wt + \pi/2)$

$= V_0 \sin(wt + \pi/2)$

The maximum value of a sine function is unity hence:

$$X_L = \frac{V_0}{I_0} = \frac{wLI_0}{I_0} = wL$$

$$\therefore\ X_L = wL = 2\pi FL$$

Since $\frac{V_0}{I_0} = \frac{V_{R.M.S.}}{I_{R.M.S.}}$ hence: $V_{R.M.S.} = X_L\, I_{R.M.S.}$

N.B. 1 When the frequency is zero, $X_L = 2\pi fL$ is also zero showing that there is no inductive effect with direct current. Inductive effects in D.C. are only important on switch on and switch off when the current is changing.

2 The reactance depends on:

a Frequency – directly proportional. The inductive effect becomes more significant at high frequency.

b Inductance – directly proportional.

3 *Capacitors*

When an alternating potential difference is applied across the plates of a parallel plate capacitor, charge is continually flowing on and off the plates. We could regard this somewhat loosely as an alternating current 'flowing' across the plates even though there is no direct conducting path for D.C. and charges never actually cross the gap.

Consider an alternating potential difference being applied across the plates.

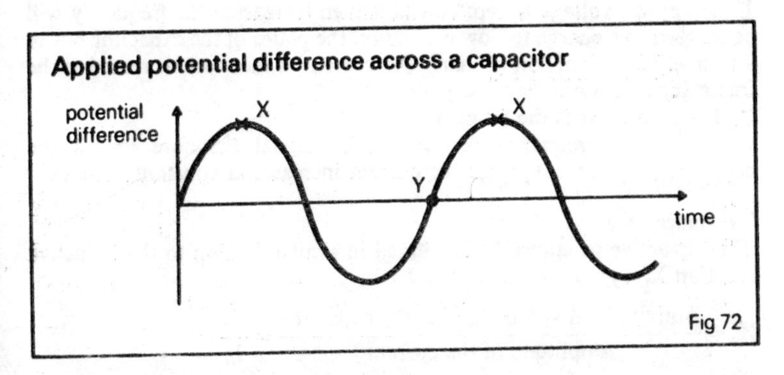

Fig 72

At the points marked X the potential difference is a maximum but is momentarily constant ($dV/dt = 0$). The current is therefore zero since there is no charge flowing either on or off the plates. At Y the applied potential difference is zero but is increasing positively at the maximum rate (steepest gradient). The rate of flow of charge (i.e. current) is therefore a maximum in the positive direction. The current and voltage are, therefore, out of phase with each other (as with the inductor) and the phase relationship is shown in Fig. 73.

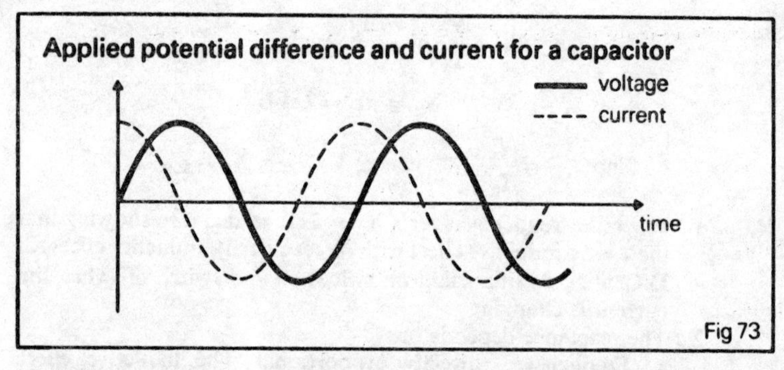

Fig 73

The voltage and current are, therefore, once again out of phase by 90° ($\pi/2$ radians) with the current leading on this occasion since it reaches its maximum first in time.

Capacitor I leads V

The behaviour of current and voltage in inductors and capacitors can, perhaps, best be memorised by remembering the word CIVIL.

CIVIL – **Capacitor cI before V**
Inductor (L) V before I

Changes of frequency

If the applied voltage is kept constant then increasing the frequency will cause the same charge to flow on and off the plates of the capacitor but in less time. The rate of flow of charge (i.e. current) is therefore increased as the frequency increases.

$\therefore I \propto f$ where f is the frequency

Increasing the frequency of the applied signal therefore lowers the capacitive reactance X_c since the current increases at constant voltage.

Reactance (X_c)

The capacitive reactance X_c is defined in similar fashion to the inductive reaction X_L by:

$$X_c = \frac{\text{Amplitude of voltage across the capacitor}}{\text{Amplitude of the current}} = \frac{V_0}{I_0}$$

Since $$V = V_0 \sin wt$$

$$\therefore Q = VC = CV_0 \sin wt$$

Current is the rate of flow of change = dQ/dt

$$\therefore I = \frac{dQ}{dt} = \frac{d}{dt}(CV_0 \sin wt) = CV_0 w \cos wt = CV_0 w \sin\left(wt + \frac{\pi}{2}\right)$$

The current can therefore be seen to be leading the voltage by $\pi/2$ and the maximum current value (I_0) is given by WCV_0 since $I = I_0 \sin(wt + \pi/2)$.

Since the maximum value of the sine function is unity we can write:

$$X_c = \frac{V_0}{I_0} = \frac{V_0}{wCV_0} = 1/wC$$

$$\therefore X_c = \frac{1}{wC} = \frac{1}{2\pi fC}$$

Since

$$\frac{V_0}{I_0} = \frac{V_{R.M.S.}}{I_{R.M.S.}}$$

hence

$$V_{R.M.S.} = X_C I_{R.M.S.}$$

N.B. 1 When f = zero X_c is infinitely large indicating that capacitors block D.C. When a D.C. voltage is applied across a capacitor a current will flow only until the charging process is complete. The charging of the capacitor is exponential as discussed in chapter 6.

2 The reactance depends on:

a Frequency – inversely proportional. The capacitance reactance is reduced as the frequency is increased.

b Capacitance – inversely proportional.

4 *Inductor and resistor in series*

This analysis is valid either for a pure inductor with a series resistor added or for a practical inductor which would necessarily possess some resistance associated with its windings.

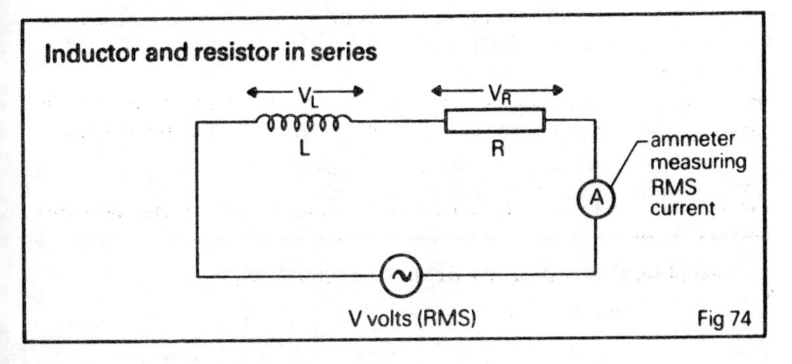

Fig 74

Impedance of the circuit
Unit Ohms (Ω) Symbol Z

In any circuit such as that shown in Fig. 74 involving resistors, inductors and capacitors it is found that the R.M.S. voltage and R.M.S. current are proportional to each other. This is usually expressed as:

$$\frac{V}{I} = Z$$

where Z is the impedance measured in ohms since it is the ratio of a voltage and current

N.B. 1 The impedance will in general possess resistive and reactive components. If there is no significant resistance present then the impedance would be purely reactive whilst if the circuit consisted of a pure resistance the impedance would be purely resistive.

N.B. 2 The analogy with Ohm's law ($V = IR$) for D.C. circuits is obvious **but** the analysis required is completely different since current and voltage are not in phase with each other for the reactive circuit elements (i.e. inductor and capacitor). Now consider specifically the circuit of Fig. 74 which shows an inductor L in series with a resistor R and an alternating current being driven by the generator. Referring to the R.M.S. values the generator is rated at V volts and the current is I amps. The potential difference across the resistor is V_R volts whilst V_L appears across the inductor.

N.B. $V \neq V_L + V_R$ (is **not** equal to)

This represents the most common mistake. The simple D.C. addition of voltages is **not** applicable here since although V_R is in phase with the current I, V_L **is not in phase with I** and we require vector addition of the voltages rather than the simpler scalar addition.

$$V_R = IR$$
$$V_L = IwL$$

and is in phase with I
and leads I and hence V_R by 90°
(since $V_L = X_L I$ and $X_L = wL$)

Representation using vectors

A direction (usually horizontal) is chosen for the current (I) and the potential difference across the inductor (V_L) and across the resistor (V_R) can then be represented on the vector diagram.

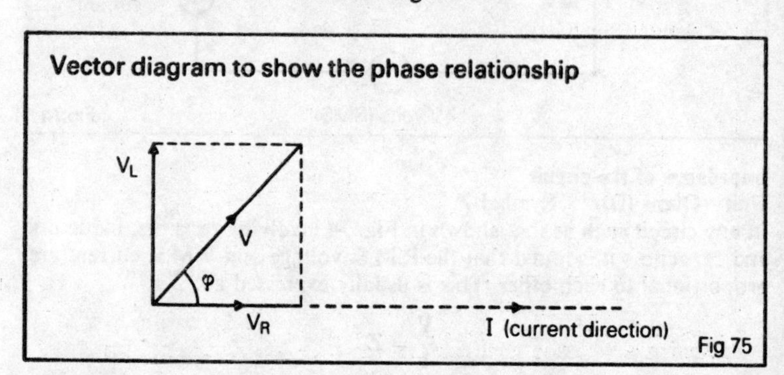

Fig 75

V_R is in phase with I whilst for the inductor V_L leads I by 90°.

The vector addition is achieved either by scale drawing or more conventionally by the use of the theorem of Pythagoras and shows the resultant voltage V leading the current by some phase angle ϕ.

$$V^2 = V_R{}^2 + V_L{}^2$$

$$\therefore \; V = (V_R{}^2 + V_L{}^2)^{1/2}$$

Since $V_R = IR$ and $V_L = IwL$ we can write:

$$V = I(R^2 + w^2L^2)^{1/2}$$

Hence the impedance $Z = \dfrac{V}{I} = (R^2 + w^2L^2)^{1/2}$

Thus V leads I by some angle where $\phi \neq 90°$.
The phase angle is given by the expression:

$$\tan\phi = \frac{V_L}{V_R} = \frac{IwL}{IR} = \frac{wL}{R}$$

Summary of inductor and resistor in series
Impedance $= Z = (R^2 + w^2L^2)^{1/2}$
Phase angle $= \phi = \tan^{-1}(wL/R)$

N.B. 1 The value of the phase angle is governed by the ratio of the resistance to the inductive reactance. When the resistance is negligible ϕ tends to 90° and the circuit behaves as a pure inductor. When the resistance is very large ϕ tends to zero as current and voltage move into phase with each other and the circuit behaves as a pure resistor.

2 Since the impedance (Z) is proportional to frequency the circuit tends to block high frequencies but pass low ones and has some use as a low pass filter.

5 *Capacitor and resistor in series*

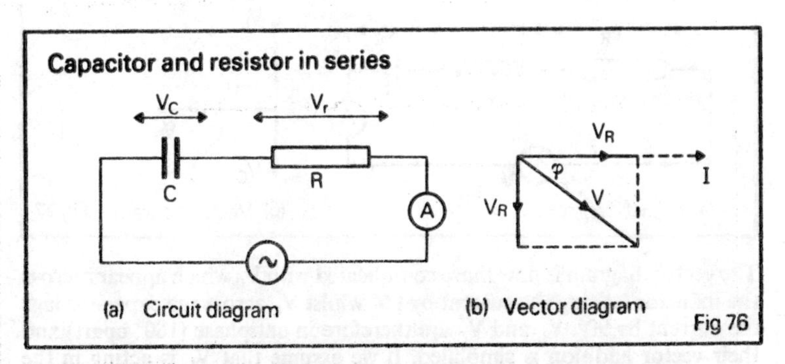

(a) Circuit diagram (b) Vector diagram Fig 76

A similar analysis to that used for the inductor is required using

$$V_C = X_C I = \frac{1}{wC} - I = \frac{I}{wC}$$

N.B. 1 The phase is lagging I by 90° for the capacitor.

2 The phase angle ϕ is measured with respect to the current direction as shown in Fig. 76.

$$V = (V_R{}^2 + V_C{}^2)^{1/2} \quad \text{(theorem of Pythagoras)}$$

Since $V_r = IR$ and $V_C = I/wC$ we can write:

$$V = I(R^2 + (1/wC)^2)^{1/2}$$

$$\therefore\ Z = V/I = (R_2 + (1/wC)^2)^{1/2}$$

$$\tan\phi = V_C/V_R = \frac{1/wC}{R} = \frac{1}{wCR}$$

Summary for capacitor and resistor in series

Impedance $= Z = (R^2 + (1/wC)^2)^{1/2}$

Phase angle $= \phi = \tan^{-1}(1/wCR)$

N.B. 1 When w = 0 (i.e. direct current) the impedance is infinite indicating that the capacitor blocks D.C. This property is used in amplifiers where a 'blocking' capacitor is inserted between one amplification stage and the next so that the D.C. conditions in one part of the circuit will not influence the operating conditions elsewhere.

2 As the frequency increases the impedance decreases and the circuit can be used as a high pass filter.

6 *Capacitor, inductor and resistor in series*

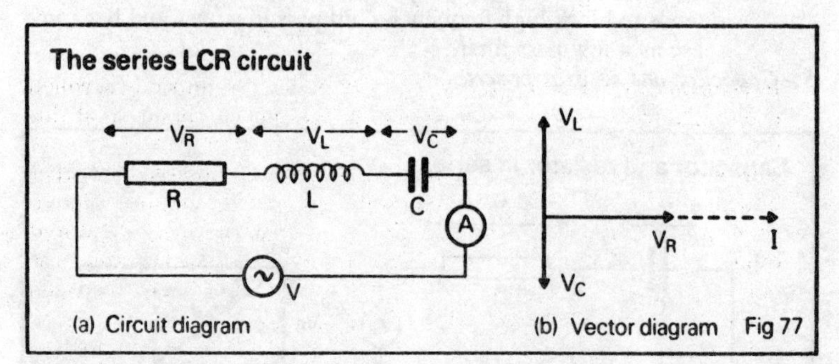

(a) Circuit diagram (b) Vector diagram Fig 77

The vector diagram is now more complicated with V_L which appears across the inductor leading the current by 90° whilst V_C across the capacitor lags the current by 90°. V_L and V_C are therefore in antiphase (180° apart) and their vector addition is simplified. If we assume that V_L is acting in the positive direction since it has been drawn as the larger contribution then the vector addition of V_L and V_C is $V_L + (-V_C)$ which simplifies to $V_L - V_C$. The application of the theorem of Pythagoras then produces:

$$V = (VR^2 + (V_L - V_C)^2)^{1/2} \quad \text{(assuming } V_L > V_C\text{)}$$

Since $V_R = IR$, $V_L = IwL$ and $V_C = I/wC$ we can write:

$$V = I(R^2 + (wL - 1/wC)^2)^{1/2}$$

and

$$Z = \frac{V}{I} = (R^2 + (wL - 1/wC)^2)^{1/2}$$

$$\tan\phi = \frac{V_L - V_C}{V_R} = \frac{I(wL - 1/wC)}{IR} = \frac{wL - 1/wC}{R}$$

Summary for the LCR circuit

$$\textbf{Impedance} = Z = (R^2 + (wL - 1/wC)^2)^{1/2}$$

$$\textbf{Phase angle} = \phi = \tan^{-1}\frac{wL - 1/wC}{R}$$

N.B. The following important points emerge from the above mathematics treatment:

a In the special case when $wL = 1/wC$ then $Z = R$ and the impedance is purely resistive and has a minimum value, the phase angle is zero indicating that the voltage is in phase with the current as would be expected in a resistive arrangement.

b When $wL = 1/wC$ the circuit is said to be in **resonance.** The resonant frequency is given by $w^2 = 1/LC$ or $w = (1/LC)^{1/2}$ but since w is the angular frequency this is, perhaps, better expressed as:

$$\textbf{Resonant frequency} = f = \frac{1}{2\pi\sqrt{LC}}$$

If the resistance R is small and $wL = 1/wC$ then the impedance which is purely resistive will be small and large currents can build up at this resonant frequency.

c This electrical resonance is similar to resonances observed in other branches of physics (e.g. mechanics, sound) in that the circuit is able to act as an energy store. At some stage of the cycle the energy is stored totally in the electric field of the capacitor whilst at another stage it is bound up in the magnetic fields (compare to the transfer between potential and kinetic energy in a simple pendulum).
Once the oscillation is established the generator is only required to make good the losses caused by the resistance. If the resistance could be eliminated the generator could, in principle, be disconnected and the energy would continue to oscillate between inductor and capacitor indefinitely.

d If a range of frequencies is applied to the circuit the only one which builds up to large values is the resonant frequency given by $w^2 = 1/LC$.

The circuit is, therefore, tuned to a particular frequency and rejects the rest.
A circuit of this type can therefore form the basis of a radio receiver. The tuned frequency can be altered by changing the value of either the inductor or capacitor and in practice a variable component is used so that the frequency can be constantly changed.
In practice a range of frequencies close to the resonant frequency will build to large values and the quality of the response is governed by the value of the resistance in the circuit.

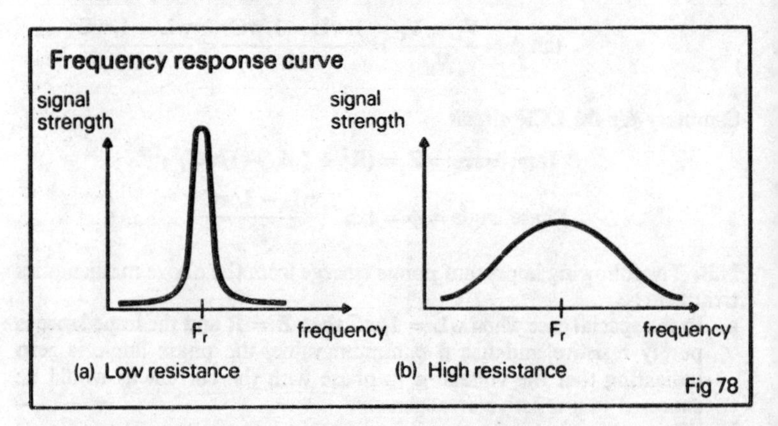

Fig 78

The resonant frequency (fr) is not altered by the resistance value but the sharpness of the response changes as shown in Fig. 78. The limiting factor in a practical circuit tends to be the design of the inductor where the inherent resistance can be reduced but never eliminated.

Power in A.C. circuits

Consider a general circuit including reactive elements so that the current (I) and voltage (V) are not in phase but are separated by some phase difference ϕ.
If $I = I_0 \sin wt$ and $V = V_0 \sin(wt + \phi)$ then the power P at any particular time (the instantaneous power) is given by:

$$P = VI = V_0 \sin wt \sin(wt + \phi)$$

Using the standard mathematical formula:
$\sin A \sin B = \frac{1}{2}[\cos(B - A) - \cos(B + A)]$ the power expression can be rewritten as:

$$P = \tfrac{1}{2}V_0I_0(\cos\phi - \cos(2wt + \phi))$$

The mean power $\bar{P}$ is the mean of the right hand side of this expression. Whilst $\cos\phi$ is constant (since ϕ is a constant phase difference) the term in $\cos(2wt + \phi)$ is an ordinary variable cosine function and therefore has a

mean value of zero. The expression therefore simplifies to:
$\bar{P} = \frac{1}{2}V_0I_0 \cos\phi = V_{R.M.S.} I_{R.M.S.} \cos\phi$

$$\textbf{Average power} = V_{R.M.S.} I_{R.M.S.} \cos\phi$$

The product $V_{R.M.S.} I_{R.M.S.}$ is called the **apparent power** and is only equal to the true power P when $\phi = 0$ and current and voltage are in phase. The term in $\cos\phi$ is known as the **power factor**.

In the special case of $\phi = 90°$ the power factor is zero, the mean power taken from the supply is also zero and the current and voltage are said to be in **quadrature**.

Plotting the current and voltage to show their 90° phase difference and then working out the power supplied at different times in the cycle shows that sometimes the power is positive and sometimes negative. This means that the power source supplies energy at some parts of the cycle but receives it back again in other parts and this situation, in fact, reverses every one quarter of one cycle.

Although no power is taken, large amounts of energy can be flowing and are stored in a reactive circuit component (inductor or capacitor) to be returned to the generator later in the cycle).

Worked examples on chapter 8

1 A pure inductor of 0·5 H is connected to a 240 volt, 50 Hz mains supply. Find the current which will flow through it.
Reactance $= X_L = wL = 2\pi fL = 2\pi \times 50 \times 0{\cdot}5 = 157\,\Omega$

$$\therefore\ I = \frac{V}{X_L} = \frac{240}{157} = 1{\cdot}53\text{ A}$$

2 Repeat the above calculation assuming that the inductor has a resistance 5 Ω and is placed in series with a 95 Ω resistor.
Reactance $= X_L = wL = 2\pi fL = 157\,\Omega$ (as before)
Total resistance $= 5 + 95 = 100\,\Omega$
Impedance $= Z = (R^2 + w^2L^2)^{1/2} = (R^2 + X_L{}^2)^{1/2}$
$\therefore\ Z = ((100)^2 + (157)^2)^{1/2} = 186\,\Omega$

$$\therefore\ I = \frac{V}{Z} = \frac{240}{186} = 1{\cdot}29\text{ A}$$

3 A 5 μf capacitor and 500 ohm resistor are connected in series with a 240 volt 50 Hz supply. Calculate **a** the current flowing and **b** the phase angle between current and voltage.

$$\text{Reactance} = X_C = \frac{1}{wC} = \frac{1}{2fC} = \frac{1}{2\pi \times 50 \times 5 \times 10^{-6}} = 637\,\Omega$$

Impedance $= Z = (R^2 + (1/wC)^2)^{1/2} = (R^2 + X_C{}^2)^{1/2}$
$\therefore\ Z = ((500)^2 + (637)^2)^{1/2} = 810\,\Omega$

$$\therefore\ I = \frac{V}{Z} = \frac{240}{810} = 0{\cdot}3\text{ A}$$

$$\tan \phi = \frac{X_C}{R} = \frac{1}{WCR} = \frac{637}{500} = 1{\cdot}27$$

Phase angle ≈ 50° with current leading.

4 An inductor and capacitor are to be joined in series to form a circuit which will resonate at 720 kHz. What inductance value is required if the capacitor has a value of 270 pf? (270×10^{-12} F.)
The resonant frequency is given by $W^2 = 1/LC$

$$\therefore\ 4\pi^2F^2 = \frac{1}{LC} \quad \text{and} \quad L = \frac{1}{4\pi^2F^2C}$$

$$\therefore\ L = \frac{1}{4\pi^2 \times 720^2 \times 10^6 \times 270 \times 10^{-12}} = 181\ \mu H$$

(In practice some resistance will also be present but this will not alter the resonant frequency.)

Practice questions

1 A 10 mH inductor has a resistance of 5 ohms and is connected to a 6 volt, 200 Hz A.C. supply. Calculate the current which will flow.

2 A 2 μF capacitor is connected in series with a 1000 ohm resistor across a 240 volt (50 Hz) A.C. supply. Calculate **a**, the current which flows **b**, the potential difference across the capacitor.

3 A 2 μF capacitor, 0·25 mH inductor and 5 ohm resistor are connected in series. Calculate **a**, the impedance at a frequency of 200 Hz, **b**, the resonant frequency of the circuit.

4 A 240 volt (R.M.S.) mains input is applied to the Y plates of a cathode ray oscilloscope with the time base turned off. If the input sensitivity is 100 volt cm^{-1}, what is the length of the line which is observed on the screen?

9 ELECTROLYSIS

The introduction in chapter 1 mentioned that there are two main types of conductor metals and electrolytes. All of the discussion so far has been concerned with metals which are undoubtedly the main type of conductor and we now turn briefly to the electrolytes.

Terminology

Electrolysis is the passage of an electric current through a solution or molten solid causing chemical changes at the electrodes.
The electrolyte is the solution (or molten solid) which undergoes electrolysis. The electrolyte must be dissociated into oppositely charged ions before electrolysis can proceed.
The electrodes are the plates through which the current enters and leaves the electrolyte. The nature of the electrodes has an important bearing on the products of the electrolysis.
The cathode is the electrode connected to the negative pole of the battery causing current flow. Positive ions in the solution travel towards the cathode and the current therefore leaves the electrolyte at this electrode.
The anode is connected to the positive pole of the battery, attracts negative ions and is the electrode by which the current enters the electrolyte.
The cations are the positive ions (usually metals or hydrogen).
The anions are the negative ions (usually radicals such as hydroxide (OH^-) or sulphate (SO_4^{2-})).

Consider the electrolysis of copper sulphate solution with carbon electrodes as one specific example of electrolysis. The ions present in the solution are $Cu(H_2O)_4^{2+}$ and SO_4^{2-} from the copper sulphate and H_3O^+ and OH^- from the water.

At the cathode $Cu(H_2O)_4^{2+}$ and H_3O^+ both arrive but the former is discharged in preference and copper is deposited on the cathode. At the anode OH^- and SO_4^{2-} both arrive but the former is discharged causing oxygen to be evolved.

The formation of ions

Consider again the structure of the atom which was discussed briefly in chapter 1. The characteristics of a particular element are determined largely by the number of protons in the nucleus and, therefore, the number of orbiting electrons because the two are equal in an atom since it is uncharged. The orbiting electrons are split into different groups or shells and do not all orbit the nucleus at the same mean separation. The innermost shell (the K shell), for example, can contain only two electrons whilst the next shell (the L shell) can contain up to 8 electrons.

Consider sodium chloride which is one of the most common ionic crystalline solids. Sodium passes eleven electrons arranged two in the K shell, eight in the L shell and the remaining one in the M shell. Chlorine possesses seventeen electrons arranged two in the K shell, eight in the L shell and seven in the M shell which is in fact one electron short of being full. The single outer electron in sodium is only weakly attracted to the nucleus since it is shielded from the attraction of eleven protons by ten electrons. The outer electrons in chlorine are also shielded from their nucleus but not to the same extent since ten electrons are now shielding seventeen protons.

When a sodium atom and a chlorine atom are in close proximity the weakly attracted eleventh sodium electron is more strongly attracted by the chlorine nucleus than the sodium nucleus and it can be regarded as joining the seven outer chlorine electrons. This rearrangement leaves a **sodium ion** (eleven protons, ten electrons) with an overall positive charge and a **chloride ion** (seventeen protons, eighteen electrons) with an overall negative charge and the two are therefore attracted together by their opposite charges.

Sodium chloride in its solid crystalline structure is therefore thought to consist of Na^+ and Cl^- ions strongly attracted to each other by the electrostatic force.

When the crystal is dropped into water this force is greatly reduced, the relative permittivity (ε_r) (chapter 5) of water is high (in fact $\varepsilon r \approx 80$ for water so the force is reduced by a factor of about eighty).

The crystal therefore dissolves and falls apart into two separate ions (i.e. it ionises) since the force holding the ions together is weakened by the solvent (water). This splitting of the ions is termed **dissociation**.

In electrolysis therefore, **mobile ions are already present in the electrolyte** and the establishment of an electric field across the electrolyte merely makes the ions move to their respective electrodes.

Faraday's laws of electrolysis

First law The mass of any substance liberated in electrolysis is proportional to the quantity of electric charge passed.

Second law This law is, perhaps, best approached by reference to an example.

The masses of the atoms of hydrogen, silver and aluminium are in the ratio 1 : 108 : 27 yet when electrolytic cells which will set free hydrogen, silver and aluminium are arranged in series so that they pass the same current it is observed that the masses liberated are in the ratio 1 : 108 : 9. Since hydrogen only possesses one electron the hydrogen ion must be singly charged and these results therefore suggest that the silver ion is also singly charged whilst the aluminium ion must carry a treble charge (Al^{3+}).

This idea that not all ions are single charged forms the basis of Faraday's second law.

The mole concept

Since a carbon atom is approximately twelve times heavier than a hydrogen atom, the numbers twelve and one could be considered significant for carbon and hydrogen respectively. 0·012 kg (12 g) of carbon is called the **relative atomic mass or one mole** of carbon and does in fact contain $6{\cdot}02 \times 10^{23}$ atoms. Similarly $6{\cdot}02 \times 10^{23}$ atoms of hydrogen constitute one mole of hydrogen and have a mass of 1 g.

This number $6{\cdot}02 \times 10^{23}$ is called the Avogadro number and is the number of atoms in a mole of any substance. Referring to the previous example, 1 mole of silver atoms have a mass of 108 g and 1 mole of aluminium atoms have a mass of 27 g.

The Faraday constant (F)

This is the charge required to liberate by electrolysis one mole of an element which forms singly charged ions (a monovalent element). The value of the constant is approximately 96,500 coulombs.

Hence one Faraday of charge (96,500 coulombs) will liberate 1 g of hydrogen, 108 g of silver but only 9 g of aluminium because the aluminium is trebly charged. The liberation of one mole of aluminium would therefore require three faradays of charge to be passed.

Worked examples on chapter 9

1 Copper has an atomic weight of 63·6 and forms ions which are doubly charged (i.e. its valency is two). How long will it take to deposit 1 kg of copper in the electrolysis of a suitable solution if the current flowing is **a** 5 amps, **b** 10 amps. (Faraday constant = 96,500 coulombs.)

Since the ion is doubly charged, two Faradays will be required to liberate 63·5 g of copper

$\therefore$ 63·5 g require $96{,}500 \times 2 = 193{,}000$ C

$\therefore$ 1 kg requires $(193{,}000 \times 1000/63{\cdot}5)$ coulombs since the amount deposited depends on the charge passed

$\therefore$ 1 kg requires $3{\cdot}04 \times 10^6$ coulombs

a Since $Q = It$, a current of 5 amps provides 5 coulombs sec^{-1}

$$\therefore \text{Time} = \frac{3{\cdot}04 \times 10^6}{5 \times 3600} = 168{\cdot}9 \text{ hours}$$

b Doubling the current to 10 amps will clearly halve the time needed since 10 coulombs sec^{-1} are now being supplied but in practice the current cannot be increased indefinitely.

Practice questions

1 A constant current is passed through copper II sulphate solution for 100 minutes and 6 g of copper are deposited on the cathode. What is the current?
(Faraday constant = 96,500 coulombs. Copper has an atomic weight of 63·6 and forms doubly charged ions.)

2 Discuss Faraday's laws of electrolysis and relate them to the ideas of ionic theory.

3 In a cell to deposit copper from copper II sulphate by electrolysis, the electrodes are 1 metre square and 10 cm apart. If the resistivity of the solution is $1{\cdot}0 \times 10^{-2}$ ohm metre, what potential difference will be required between the electrodes if the deposition rate is to be 0·5 kg per hour. (See question one for relevant data on copper.)

10 SEMICONDUCTORS

Atomic spectra

When a gas at reduced pressure is excited in a discharge tube radiation is emitted and observed to contain certain wavelengths only. This phenomenon of atomic spectra is best studied by observing the discharge either with a prism or diffraction grating, both of which will split the radiation into its constituent wavelengths.

Every element displays a unique line spectrum when a sample of it in the vapour phase is excited to produce an **emission spectrum**. The analysis of an unknown substance by comparing its spectrum to known spectra is part of the branch of physics called **spectroscopy**.

The existence of line spectra is difficult to explain in terms of the so called 'classical' theory but the **quantum theory** is fairly successful.

The Bohr model of the hydrogen atom

A full discussion of the quantum theory is beyond the scope of this revision guide but the Bohr model for hydrogen can be used to explain spectral lines.

1 Electrons move around the nucleus in different orbits (K shell, L shell, etc – chapter 9) but in a stable orbit no energy is radiated by the electron.
2 Only certain orbits are allowed.
3 Each allowed orbit has a particular energy associated with it. An electron can, therefore, only possess certain discrete energy values. Most of the theoretically possible energies are in fact forbidden and the energy is said to be quantised.
4 If an orbit has an energy E_1 associated with it and the next possible orbit has an energy of E_2 ($E_2 > E_1$) then an electron cannot possess an energy value in between E_1 and E_2. Should the electron change from the E_2 orbit to the E_1 orbit then an amount of energy ($E_2 - E_1$) will be given out and we can write:

$$E_2 - E_1 = hf$$

where h is Planck's constant and f is the frequency at which the energy is released.
5 In a discharge tube electrons accept a definite amount of energy and rise to higher levels as they go into a **excited state**. They then lose this extra energy in falling back to their **ground state** (stable state) and the energy is radiated at a particular frequency.
6 Different electron transitions between the possible energy levels cause different amounts of energy to be radiated. Each amount of energy is radiated at a single frequency. The observed spectrum therefore contains a number of discrete frequencies corresponding to the different possible transitions between energy levels.

Band spectra

The spectra described so far are all line spectra but when a molecular gas (e.g. oxygen) is excited, the spectrum becomes more complicated and is divided into a series of bands.

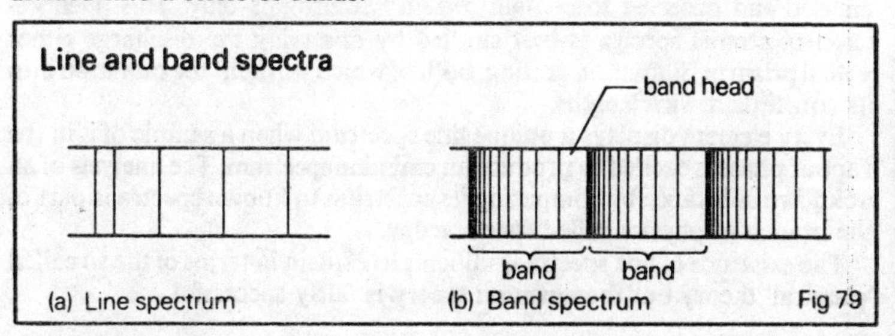

(a) Line spectrum (b) Band spectrum Fig 79

The band head is usually fairly well defined but the band 'fades' at the other end. This spacing of the bands is sometimes termed fluting.

The band theory of solids

In a solid where there are many atoms in close proximity, the spectrum becomes even more complicated and the energy bands contain energy levels which are very close to each other. The bands may or may not overlap depending on the type of material.

a **Conductors** The energy bands do overlap and all electron energy levels are allowed. An electron can assume any energy value and is free to move throughout the solid.

b **Semiconductors** The energy bands almost overlap but there is a **band gap** or **forbidden energy gap** between two bands. The electrons require a little energy to enable them to jump the band gap and move to the next allowed band. The semiconductor will not conduct of its own accord but can be made to do so fairly easily.

c **Insulators** The band gap is now so large that the bands are well separated and there is no easy movement of electrons. No conductance of electricity is possible under normal circumstances though the insulation could be broken if an excessively high voltage were applied.

Valence and conduction band

The **valence band** is the lowest energy band available whilst the **conductance band** is the next highest energy band. Electrons must be in the conduction band before conduction can take place.

In an insulator the valence band is full of electrons but the conduction band is empty and electrons cannot cross the band gap.

A semiconductor has a full valence band and empty conduction band at absolute zero (OK) and is therefore an insulator but at normal temperatures some electrons can acquire enough thermal energy to jump the band gap and reach the conduction band.

The conductors (metals) possess overlapping bands and electrons can, therefore, always occupy the conduction band.

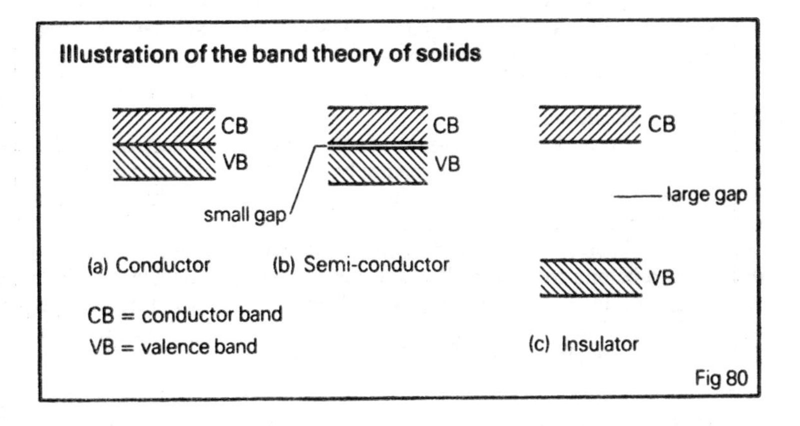

Fig 80

Conduction mechanisms in semiconductors

1 *Intrinsic*

Consider raising the temperature of a semiconductor which is initially at absolute zero. The following points should be noted:

a At absolute zero all of the outer electrons are used in bonding the solid together and none are free to carry an electrical current. This is described on band theory by saying that the valence band is full whilst the conduction band is empty and no electron is able to jump the band gap which exists between the two.

b As the temperature is increased, thermal vibrations increase and at room temperature, for example, a few electrons have enough energy to break away from their parent atoms. On band theory they have enough energy to jump the band gap and enter the conduction band.

c Hence **as the temperature increases the electrical conductivity** increases as more electrons are excited into the conduction band. **Metals and semiconductors therefore show opposite behaviour to temperature increase** since in a metal the resistance increases as the temperature is raised.

d When an electron escapes from the previously full valence band it leaves behind a positive 'hole' (i.e. the absence of an electron).

e Other electrons in the valence band find it energetically favourable to fill the hole (i.e. combine with it) but this process will only leave another hole elsewhere since another part of the valence band will now be missing one electron.

Since the hole can be regarded as moving it is often convenient to discuss the working of the semiconductor in terms of hole currents as well as electron currents. As the hole moves in one direction, electrons are moving in the opposite direction and the hole behaves very much as though it was a positively charged electron.

N.B. In the discussion so far, one electron left the valence band to go to the conduction band and created one hole. **The number of electrons in the conduction band is equal to the number of holes in the valence band and the semiconductor is said to be intrinsic.**

2 *Doped semiconductors*

In practice semiconductors do not have equal numbers of electrons in the conduction band and holes in the valence band and are said to be **extrinsic.** This unbalance is caused by deliberately **doping** the semiconductor with a selected amount of a known **impurity**.

Consider silicon and germanium which are the most common semiconductors and belong to Group IV of the periodic table of elements having four electrons in their outer shell. In silicon a stable structure can be achieved by each silicon atom bonding with four of its neighbours to form four so called covalent bonds (Fig. 81a).

Structure of silicon

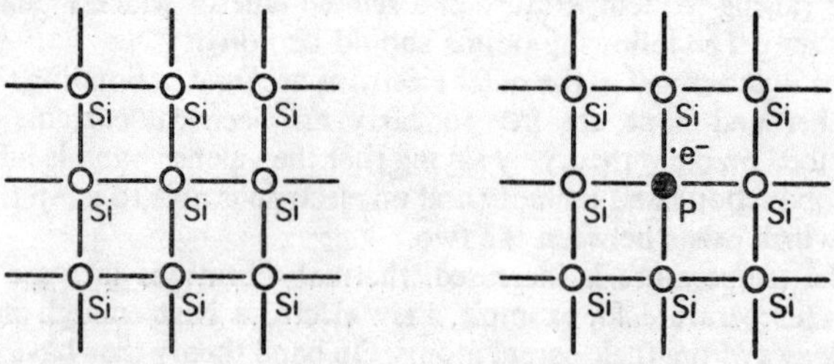

(a) Pure silicon

(b) Doped silicon

Fig 81

When an impurity from group V (five electrons in the outer shell) such as phosphorus (arsenic or antimony) is introduced, it will fit into the structure as shown in Fig. 81b but there is one electron left over which is only loosely attached to the impurity atom. This extra electron is easily detached by thermal vibrations and can move into the conduction band but **does not leave a hole behind** in the valence bond.

The phosphorus has donated an electron to the conduction band and is said to be a **donor impurity**. The extra electron is said to occupy a **donor level** which is drawn just below the conduction band on band theory to show that access to the conduction band is easily achievable.

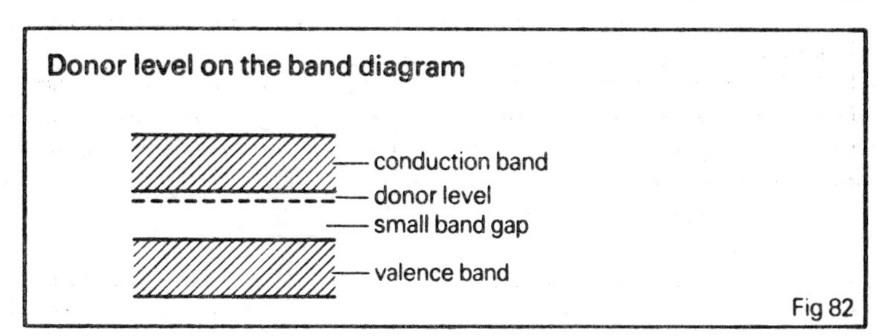

Fig 82

Since there are now more electrons in the conduction band than holes in the valence band, the semiconductor is said to be **n type**. This does **not** mean that the semiconductor is negatively charged but expresses the idea that the conduction is mainly by negative charge carriers.

In an n type semiconductor electrons are the **majority carriers** but there are still some holes present due to intrinsic conductivity and these are called the **minority carriers**.

By a similar argument, when an impurity from Group III (three electrons in the outer shell) such as gallium or indium is introduced, excess holes are created in the valence band. The impurity is called an **acceptor impurity** and creates acceptor levels above the valence band. The semiconductor is then termed **p type**.

The pn–junction diode

A diode is a device which will conduct electricity in one direction only. It can therefore be used to rectify alternating current to direct as discussed in chapter 8.

The silicon junction diode consists of a single crystal of pure silicon which has been made n type in one region and p type in another by adding different impurities. A junction is therefore formed between the p and n type regions. **N.B.** It is important to start with one crystal. It is not possible to make separate p type and n type crystals and then join them later since the conduction is spoilt by defects at the interface.

The n type possesses excess electrons in the conduction band whilst the p type has excess holes in the valence band but this implies **neither** that the n type is negative **nor** that the p type is positive.

As soon as the junction is formed, electrons from the n type will diffuse into the p type to fill its holes and holes from the p type diffuse into the n type to be filled by electrons. A neutral zone or **depletion region** is therefore formed at the junction but the exchange of charge carriers is only momentary since a potential barrier is established preventing further current flow.

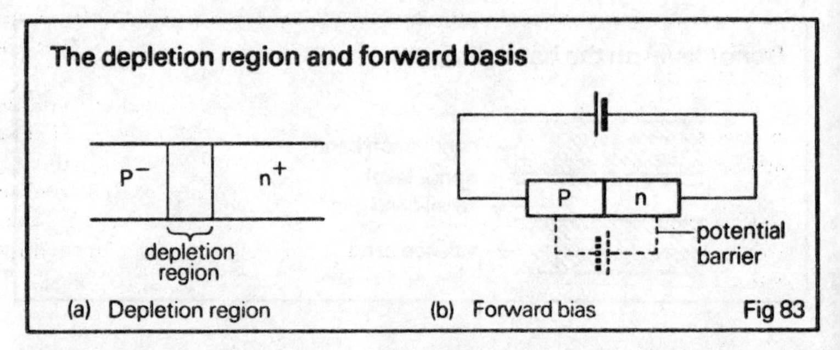

(a) Depletion region (b) Forward bias Fig 83

Since the n type is losing electrons it charges positively whilst the p type charges negatively and the potential barrier set up (about 0·1 volt) prevents further current flow. The rectifying properties of the device are due to this potential barrier.

Forward and reverse bias

If an external potential difference is applied across the junction in opposition to the potential barrier (Fig. 83b) then the electric field in the depletion region is reduced and a current can flow. The junction is said to be on **forward bias**. If, however, the external potential difference is reversed then the junction is on **reverse bias** and the flow of carriers across the junction is inhibited even more since the external potential difference is now increasing the size of the potential barrier.

A small current is, in fact, possible in reverse bias due to minority carriers created by intrinsic conductivity. Electrons from the p type entering the n type constitute a minority carrier current but this is normally negligible compared to the forward bias majority carrier current.

The junction therefore conducts current easily in the forward direction but has a high resistance in the reverse direction and the current/voltage characteristic is shown in Fig. 84.

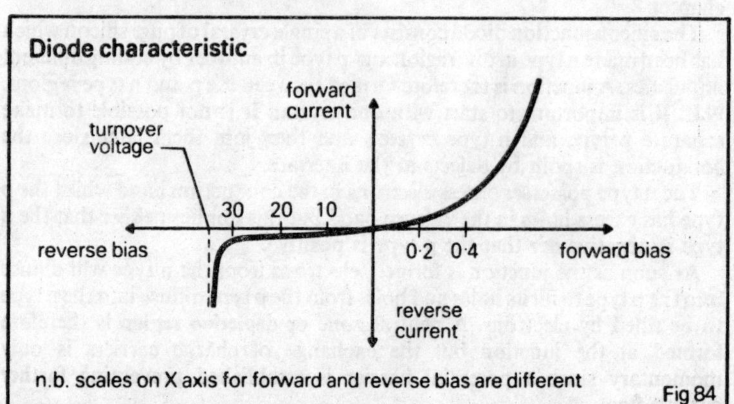

Fig 84

Ohm's law is not obeyed since the forward and reverse characteristics are different and in the forward direction the current grows exponentially as the potential difference increases.

In the reverse direction the current is negligible until the turnover voltage is reached when the barrier region breaks down because the electric field is strong enough to elevate electrons directly into the conduction band.

The rapid increase after breakdown is called a **zener or avalanche effect** and irreversible damage is likely to result unless the current is limited. Zener diodes use this effect as a way of stabilising a voltage to a particular required value.

The transistor

This is essentially two pn junctions formed back to back on a single crystal. There are two possibilities:

a **The pnp junction transistor** This consists of an n type region sandwiched between two p type regions. The n type central region is called the **base.** The p type regions are called the **emitter and collector.**

b **The npn junction transistor** The p type base is sandwiched between two n type regions.

Action of the transistor

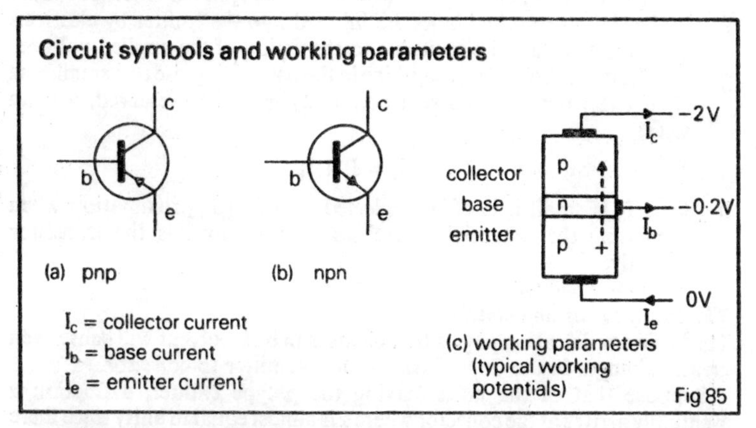

Fig 85

As usual, the arrows show the direction of conventional current flow (i.e. hole flow). Here with reference to Fig. 85c we choose to discuss the pnp transistor and consider its operation in terms of the flow of holes (remembering that the hole can be regarded as a moving positive charge).

When the transistor is first fabricated two depletion regions will be established since the device is essentially two pn junctions back to back. The base/emitter junction is forward biased as can be seen with reference to Fig. 85c and 83 whilst the base collector junction is reverse biased.

Base/emitter junction – forward bias
Base/collector junction – reverse bias

a Since the emitter/base junction is on forward bias, charges can flow and positive holes (charge carriers in the p type emitter) can cross the junction into the base since the applied external electric field overcomes the potential barrier.

b Electrons can also flow from base to emitter but this current can usually be ignored since the hole concentration in the p type regions is usually arranged to be much greater than the electron concentration in the n type. (i.e. the p type is far more heavily doped than the n type. The emitter and collection regions are more heavily doped than the base.)

c Most of the holes go through the n type base and only a few recombine with electrons since **the base region is very thin**. This recombination produces base current I_b which replaces the electrons which are lost.

d On reaching the base/collector junction the holes enter the collector. Since this junction is on reverse bias it prevents holes from the p type collector entering the n type base but it will, therefore, assist holes in the base trying to reach the collector. In a normal pn junction, holes in the n type would constitute a minority carrier and could be ignored but in the case of the transistor these holes have been provided by the p type emitter and are present in considerable numbers.

e Holes reaching the collector combine with circuit electrons at the collector terminal causing a collector current I_c in the external circuit.

f At the emitter terminal holes are injected into the emitter by electrons entering the external circuit due to the action of the applied voltage.

g $I_e > I_c$ since a few holes recombine in the base giving rise to a small base current (I_b) but since current continuity must be preserved, we can write:

$$\mathbf{I_e = I_c + I_b}$$

h Since I_b is small it is often neglected to a first approximation when calculating the potential differences required to bias the transistor correctly.

The transistor as an amplifier

The basic amplification idea is that changes in base current will cause even greater changes in the flow of current from emitter to collector.

Suppose that of the holes leaving the p type emitter, a fraction α eventually arrives at the collector where α is almost equal to unity since there is little recombination with electrons in the thin n type base.

$\therefore\ I_c = \alpha I_e$ where $\alpha = 0{\cdot}97$ is a typical value

$I_e = I_c + I_B$ (current continuity)

Hence replacing I_e by I_c/α we have:

$$\frac{I_c}{\alpha} = I_c + I_b \quad \text{and} \quad I_c = \alpha I_c + \alpha I_b$$

$$\therefore\ I_c(1-\alpha) = \alpha I_b \quad \text{so} \quad I_c = \left(\frac{\alpha}{1-2}\right)I_b$$

$$\therefore\ \delta I_c = \left(\frac{\alpha}{1-\alpha}\right)\delta I_b$$

where δI_c and δI_b represent small changes in collector and base currents respectively.

The current amplification factor (β) is given by:

$$\beta = \frac{\delta I_c}{\delta I_b} = \frac{\alpha}{1-2} = \frac{0{\cdot}97}{0{\cdot}03} \approx 32$$

To achieve this amplification in practice a small alternating current variation is superimposed on the D.C. bias to the base. This causes a larger alternating current variation from emitter to collector which is again superimposed on the D.C. conditions.

The input variation must be fairly small or the output variation is so large that it upsets the D.C. bias conditions and causes the output to become distorted.

The common emitter circuit

This is the most common transistor configuration and can be used to demonstrate amplification.

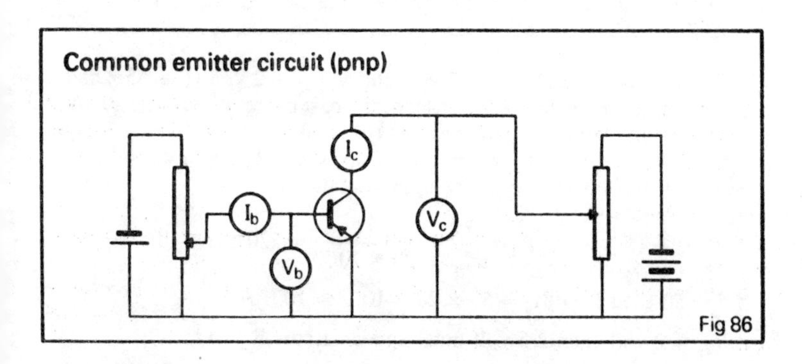

Common emitter circuit (pnp)

Fig 86

The circuit is called common emitter since it is the emitter which is connected to both the base and collector via the external circuit. The biasing conditions indicated on Fig. 85c are achieved using two separate circuits as shown in Fig. 86.

The characteristics of the device can be studied by altering the parameters two at a time by varying the rheostats.

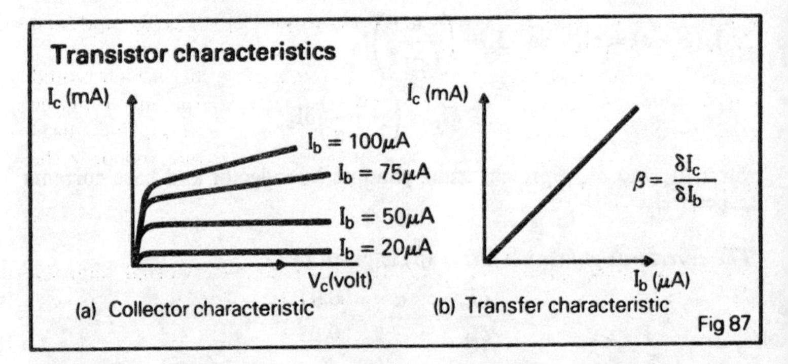

(a) Collector characteristic (b) Transfer characteristic

Fig 87

In practice the base and collector are usually given the correct bias voltages from a single battery rather than two separate circuits.

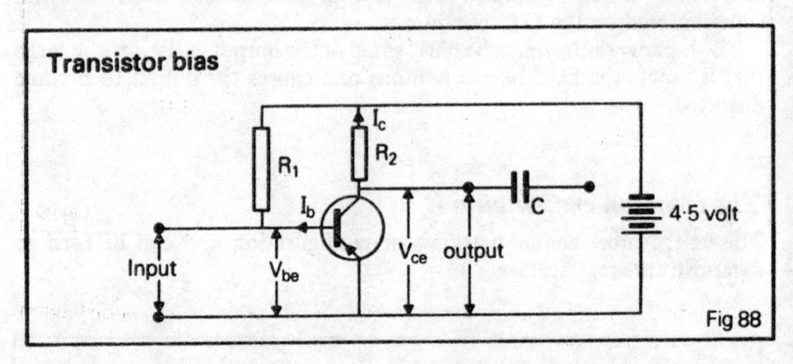

Fig 88

Suppose we require V_{ce} = 2 volt and V_{be} = 0·2 volt (Fig. 85c) and are given the transistor specifications that the base current I_b is to be 20 μA and that the current amplification factor (β) is 50.

Since $V_{be} \approx 0{\cdot}2$ volt the potential drop across R_1 will need to be $\approx$ 4·3 volt (call this 4·5 volt to a first approximation).

$$\therefore R_1 \approx \frac{4{\cdot}5}{I_b} = \frac{4{\cdot}5}{20 \times 10^{-6}} = 220\,\text{k}\,\Omega$$

If $\beta = 50$ then $I_c = \beta I_b = 50 \times 20 \times 10^{-6} = 10^{-3}$ A

If V_{ce} = 2 volt then 2·5 volt must appear across R_2 and:

$$R_2 = \frac{2{\cdot}5}{1 \times 10^{-3}} = 2{\cdot}5\,\text{K}\Omega$$

Inserting these resistors should be good enough to turn the transistor on but a check of the waveforms observed would now be needed using an oscilloscope to ensure the absence of distortion which could occur if the transistor did not meet its theoretical specification. The resistors could then be changed slightly if necessary.

Figure 88 shows only one stage of a complete amplifier which would consist of several stages **cascaded** together by connecting the output of one stage to the input of the next. A **blocking capacitor** (C) (chapter 8) is used between stages so that the D.C. component is filtered out and only the alternating part of the signal appears in the output.

The Hall effect

Consider a current moving along a slab of metal with an external magnetic field acting at right angles to the current direction (Fig. 89).

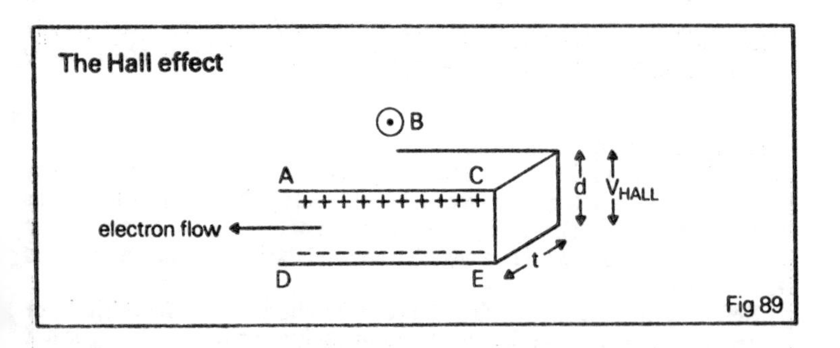

Fig 89

Since the force on a wire length ℓ carrying a current I in a magnetic field of flux density B is given by:

$$F = BI\ell = BQV$$

(chapter 2) – where Q is the charge and V is the velocity of the charge carriers

Hence each electron (of charge e) experiences a force BeV at right angles to its direction of motion and electrons will, therefore, be deflected towards the face DE. This face will charge negatively and lower its potential with respect to the upper force AC and the charging process will continue until the potential difference established is large enough to prevent any further electron flow. This potential difference between the faces AC and DE is known as the **Hall voltage** (V_{HALL}).

When equilibrium is established the electric field opposes the force produced by the magnetic field and we can write:

$$eE = \frac{eV_{HALL}}{d} = BeV$$

Hence $V_{HALL} = BVd$

Since $I = nAVe$ (drift velocity expression, chapter 1)

and $A = dt$ (where t is the thickness)

$$\therefore \; V = \frac{I}{Ndte} \quad \text{and} \quad V_{HALL} = \frac{BId}{Ndte} = \frac{BI}{net}$$

$$V_{HALL} = \frac{BI}{net}$$

Choosing reasonable values (which could be obtained in the laboratory) to maximise V_{HALL} it is found that the voltage in metals is very small. For example $B = IT$, $n = 10^{29}$ electron metre^{-3}, $e = 1{\cdot}6 \times 10^{-19}$ C, $I = 10$ A and $t \approx 1$ mm produces a Hall voltage at best of about 1 microvolt which could be measured, but not easily.

The Hall effect is not particularly important in metals but in semiconductors the number of charge carriers per metre^3 (n) is much less (typically 10^{25} per m^3) and the Hall voltage is now of order millivolts and is easily measured.

The Hall effect and semiconductors

In semiconductors the charge carriers can be positive or negative but they are still deflected to the same face of the material by the magnetic field. This is because both the charge sign and direction of motion are reversed for a positive carrier compared to a negative one and the force is, therefore, still in the same direction.

Referring to Fig. 89, the face DE can be charged either positively or negatively depending on the predominant charge carrier and V_{HALL} is not only larger but can be positive or negative.

Measuring the voltage will therefore show whether the material is p type or n type.

The Hall probe

This consists of a semiconducting wafer which has contacts on opposite sides which are connected to a high impedance voltmeter. When the probe is inserted into a magnetic field and a current is passed through it, the Hall voltage can be measured on the voltmeter.

Sinc $V_{HALL} = \frac{BI}{net}$ hence $B = \frac{V_{HALL}\, net}{I}$

The flux density of the magnetic field can be found by measuring V_{HALL} provided 'net' which is a constant for a particular semiconductor has previously been determined.

The Hall probe is an alternative to the search coil (chapter 4) for the measurement of magnetic flux density.

Practice questions

1 Discuss the differences between a conductor, semiconductor and insulator by considering the band theory of solids.

2 Explain the terms intrinsic and extrinsic semiconductors.
3 Discuss the operation of a pn junction and describe the essential features of its V/I characteristic.
4 Discuss briefly the operation of a transistor. What is meant by the term 'common emitter mode'?
5 Explain the use of a Hall probe for the measurement of magnetic flux density.

11 ANSWERS TO NUMERICAL QUESTIONS

Chapter 1

1 5·5 Ω
2 18·3 metres in parallel
3 62·7 Ω
4 $6 \times 10^{-3}\ ^{\circ}C^{-1}$

Chapter 2

1 a 0·1 Ω in parallel, b 9900 Ω in series
2 4·5 mV
3 6·7 cm from 10 A wire
4 $1{\cdot}25 \times 10^{-4}\ Nm^{-1}$

Chapter 3

3 0·63 Ω
4 $1{\cdot}89 \times 10^{-7}\ \Omega m$

Chapter 4

1 4·8 amps
2 $9{\cdot}6 \times 10^{-4}\ T$

Chapter 5

1 $4{\cdot}4 \times 10^{-8}\ cm^{-2}$, $2{\cdot}2 \times 10^{-8}\ C$, 990 V
2 $5 \times 10^{-19}\ C$
3 4·7 μC, $1{\cdot}49 \times 10^{-5}\ C$
4 $1{\cdot}2 \times 10^{6}$ volts

Chapter 6

2 4
4 75 volts, $2{\cdot}24 \times 10^{-2}\ J$, $6 \times 10^{-2}\ J$

Chapter 7

5 67 Hz

Chapter 8

1 0·44 amps
2 128 mA, 204 volts
3 37 Ω 7·0 kHz
4 6·78 cm

Chapter 9

1 3·03 A
3 0·42 volt